Network Connectivity

Concepts, Computation, and Optimization

Synthesis Lectures on Learning, Networks, and Algorithms

Editor
Lei Ying, *University of Michigan, Ann Arbor*

Editor Emeritus
R. Srikant, *University of Illinois at Urbana-Champaign*

Founding Editor Emeritus
Jean Walrand, *University of California, Berkeley*

Synthesis Lectures on Learning, Networks, and Algorithms is an ongoing series of 75- to 150-page publications on topics on the design, analysis, and management of complex networked systems using tools from control, communications, learning, optimization, and stochastic analysis. Each lecture is a self-contained presentation of one topic by a leading expert. The topics include learning, networks, and algorithms, and cover a broad spectrum of applications to networked systems including communication networks, data-center networks, social, and transportation networks. The series is designed to:

- Provide the best available presentations of important aspects of complex networked systems.

- Help engineers and advanced students keep up with recent developments in a rapidly evolving field of science and technology.

- Facilitate the development of courses in this field.

Network Simulation
Richard M. Fujimoto, Kalyan S. Perumalla, and George F. Riley
2006

Network Connectivity: Concepts, Computation, and Optimization
Chen Chen and Hanghang Tong

ISBN: 978-3-031-03756-6   paperback
ISBN: 978-3-031-03766-5   PDF
ISBN: 978-3-031-03776-4   hardcover

DOI 10.1007/978-3-031-03766-5

A Publication in the Springer series
SYNTHESIS LECTURES ON LEARNING, NETWORKS, AND ALGORITHMS

Lecture #28
Editor: Lei Ying, *University of Michigan, Ann Arbor*
Editor Emeritus: R. Srikant, *University of Illinois at Urbana-Champaign*
Founding Editor Emeritus: Jean Walrand, *University of California, Berkeley*
Series ISSN
Print 2690-4306   Electronic 2690-4314

Network Connectivity

Concepts, Computation, and Optimization

Chen Chen
University of Virginia

Hanghang Tong
University of Illinois at Urbana-Champaign

*SYNTHESIS LECTURES ON LEARNING, NETWORKS, AND ALGORITHMS
#28*

ABSTRACT

Networks naturally appear in many high-impact domains, ranging from social network analysis to disease dissemination studies to infrastructure system design. Within network studies, network connectivity plays an important role in a myriad of applications. The diversity of application areas has spurred numerous connectivity measures, each designed for some specific tasks. Depending on the complexity of connectivity measures, the computational cost of calculating the connectivity score can vary significantly. Moreover, the complexity of the connectivity would predominantly affect the hardness of connectivity optimization, which is a fundamental problem for network connectivity studies.

This book presents a thorough study in network connectivity, including its concepts, computation, and optimization. Specifically, a unified connectivity measure model will be introduced to unveil the commonality among existing connectivity measures. For the connectivity computation aspect, the authors introduce the connectivity tracking problems and present several effective connectivity inference frameworks under different network settings. Taking the connectivity optimization perspective, the book analyzes the problem theoretically and introduces an approximation framework to effectively optimize the network connectivity. Lastly, the book discusses the new research frontiers and directions to explore for network connectivity studies.

This book is an accessible introduction to the study of connectivity in complex networks. It is essential reading for advanced undergraduates, Ph.D. students, as well as researchers and practitioners who are interested in graph mining, data mining, and machine learning.

KEYWORDS

network connectivity, paths and connectivity problems, dynamic network analysis, graph mining, data mining, machine learning

To my parents and husband, for their love and support.

–CC

To Rongrong Tong, Fang Li, and Muge Li.

–HT

Contents

Acknowledgments

First, we are thankful to our colleagues in the DATA and STAR Labs at Arizona State University (current iDEA and iSAIL Labs at University of Illinois, Urbana-Champaign): Liangyue Li, Xing Su, Si Zhang, Boxin Du, Qinghai Zhou, Jian Kang, Zhe Xu, Lihui Liu, Scott Freitas, Haichao Yu, Ruiyue Peng, Rongyu Lin, Xiaoyu Zhang, Dawei Zhou, Yao Zhou, Arun Reddy, Xu Liu, Xue Hu, Jun Wu, Lecheng Zheng, Dongqi Fu, Pei Yang, and Qi Tan for their support.

We are also grateful to our collaborators (in alphabetical order): Nadya Bliss, Duen Horng Chau, Tina Eliassi-Rad, Christos Faloutsos, Michalis Faloutsos, Jingrui He, Qing He, Jundong Li, Huan Liu, B. Aditya Prakash, Charalampos E. Tsourakakis, Yinglong Xia, Lei Xie, Lei Ying, and Hui Zang for their valuable contribution on this work. Particularly, we would like to acknowledge Lei Ying and Michael Morgan for their encouragement and help throughout the preparation of this book. This work is supported by NSF (1947135) and DTRA (HDTRA1-16-0017).

Last, we would like to express our deep gratitude to our families for their unconditional love and support all the time.

Chen Chen and Hanghang Tong
December 2021

CHAPTER 1

Introduction

1.1 BACKGROUND

Networks are prevalent in many high-impact domains, including information dissemination, social collaboration, infrastructure constructions, and many more. The most well-studied type of network is the single-layered network, where the nodes are collected from the same domain and the links are used to represent the same type of connections. For example, in Figure 1.1a, we have three single-layered networks, which are the power grid network, the autonomous system network (AS network), and the transportation network. Specifically, the nodes in the power grid are the power stations and the edges are electricity transmission lines; the nodes in the AS network are the routers and the edges are packet transmission paths, while in the transportation network, the nodes can be the train stations and the edges are the railroads that connect those stations. However, as the world is becoming highly connected, cross-domain interactions are more frequently observed in numerous applications, catalyzing the emergence of a new network model—*multi-layered networks* [12, 41, 96, 108]. One typical example of such type of network is the critical infrastructure network as illustrated in Figure 1.1b. In an infrastructure network system, the full functioning of the autonomous system network (AS network) and the transportation network is dependent on the power supply from the power grid. While for the gas-fired and coal-fired generators in the power grid, their functioning is fully dependent on the gas and coal supply from the transportation network. Moreover, to keep the whole complex system working in order, extensive communications are needed between the nodes in the networks, which are supported by the AS network. In addition to the infrastructure systems, multi-layered networks also appear in many other application domains, such as organization-level collaboration platform [15] and cross-platform e-commerce systems [24, 69, 78, 138].

1.2 MOTIVATIONS

What is Network Connectivity? Network connectivity is used to measure how well different parts of the network are connected to each other, which plays a crucial role in applications like disease control, network robustness analysis, community detection, etc. Correspondingly, different connectivity measures are designed for each of the applications. Examples include epidemic threshold [13] for disease dissemination analysis, natural connectivity [51] for robustness measurement, and triangle capacity for social network mining. Empirical analysis has demonstrated the effectiveness of those connectivity measures in their own tasks, but none of them can

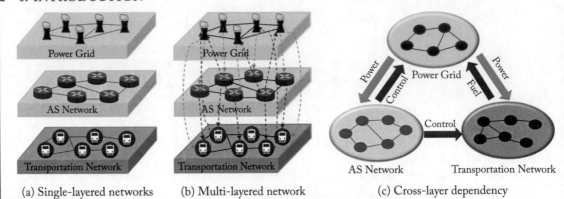

(a) Single-layered networks   (b) Multi-layered network   (c) Cross-layer dependency

Figure 1.1: Illustrative examples of networks. (a) shows the examples of single-layered networks (i.e., power grid, AS network, and transportation network). (b) is a three-layered critical infrastructure network by considering the cross-layer dependencies in (a). (c) is the abstraction of the cross-layer dependency in the critical infrastructure network. Specifically, each ellipse corresponds to a single-layered network. The arrows between two ellipses indicate cross-layer dependency relationships between the corresponding two networks (e.g., a router in the AS network depends on one or more power plants in the power grid).

be used as a common measure across different domains. Furthermore, most, if not all, of the existing connectivity measures are defined on single-layered networks, leaving the problem of measuring multi-layered network connectivity unexplored.

How to Compute Network Connectivity? Existing network connectivity research predominantly assumes that the input network is static and accurate, which does not fit into the dynamic and noisy real-world settings. Real-world networks are evolving over time. In some cases, subtle changes in the network structure may lead to huge differences on some of the connectivity measures. For example, in websites like Facebook and Twitter, new connections between users emerge all the time, which would, in turn, change the influential individuals in the network. Thus, it is crucial for online marketing companies to keep track of those changes since their advertisement targeting strategies may need to be modified accordingly. By the nature of network connectivity, its computation often involves complicated operations over global network structure (e.g., eigenvalue computation for epidemic threshold [13] and natural connectivity [51]). Thus, it would be computationally costly to recompute the connectivity measures at each time stamp for tracking purposes. To address this issue, it is necessary to develop an efficient incremental connectivity update framework that can effectively approximate most connectivity measures in the long run. On the other hand, in multi-layered networks, it remains a daunting task to know the exact cross-layer dependency structure due to noise, incomplete data source and limited accessibility issues. For example, an extreme weather event might significantly disrupt the power grid, transportation network, and cross-layer dependencies in between at the epicenter. Yet, due to limited accessibility to the damaged area during or soon after the disruption, the cross-layer dependency structure might only have a probabilistic and/or coarse-grained description. Therefore, it is critical to infer those missing dependencies before computing the connectivity in such multi-layered networks.

How to Optimize Network Connectivity? One of the most important tasks for network connectivity studies is to optimize (minimize/maximize) the connectivity score by adjusting the underlying network structure. Previous literature has proved that the optimization problem on epidemic threshold and triangle capacity on single-layered networks is NP-hard. However, for some complex connectivity measures (e.g., natural connectivity), the hardness of the corresponding optimization problems still remains unknown. Most importantly, existing connectivity optimization methods are mainly based on single-layered networks. Compared to single-layered networks, multi-layered networks are more sensitive to disturbance since its effect may be amplified through cross-layer dependencies in all the dependent networks, leading to a cascade failure of the entire system. To tackle the connectivity optimization problem in multi-layered networks, great efforts have been made from different research areas for manipulating *two-layered* interdependent network systems [12, 41, 96, 108]. Although much progress has been made, challenges are still largely open. First, as the connectivity measures are highly diversified, the ad-hoc optimization algorithms that are effective for specific measures may not work well on other measures. Thus, the problem of how to design a generic optimization strategy for a wide range of network connectivity measures is in need of investigation. Second, existing optimization strategies tailored for two-layered networks might be sub-optimal, or even misleading for arbitrarily structured multi-layered networks, because they cannot effectively unravel the nested dependency structure in the network.

1.3   RESEARCH TASKS OVERVIEW

The main problems introduced in this book are focused on the **measure concepts**, **inference computation**, and **optimization** of network connectivity in complex networks. The relationship between those problems is shown in Figure 1.2. Generally speaking, a well-defined connectivity measure serves as the objective to inference computation and optimization tasks. The inference results, in turn, provide a good approximation on the connectivity measure and improve the accuracy of the input network for optimization tasks. Last, the optimization methods are used to find optimal strategies to manipulate the network structure, which can effectively change the connectivity of the network and influence the inference results from task 2.

1.4   ORGANIZATION

The remainder of the book is organized as follows. In Chapter 2, we introduce the generalized definition for connectivity measure concepts in single-layered networks and its extension

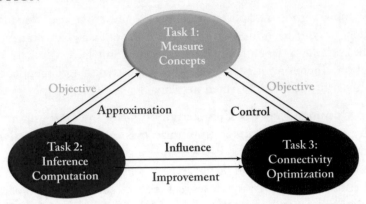

Figure 1.2: Tasks overview.

on multi-layered networks. In Chapter 3, we address the eigen-functions (connectivity) tracking problem in dynamic networks and the cross-layer dependence inference problem in multi-layered networks. In Chapter 4, we study the fundamental limits and efficient algorithms for the network connectivity optimization problem in single-layered networks and its extension on multi-layered networks. Finally, in Chapter 5, we conclude the book and discuss future directions for network connectivity studies.

CHAPTER 2

Connectivity Measure Concepts

Connectivity is a fundamental property of networks and has been a core research theme in graph theory and mining for decades.[1] At the macro-level, network connectivity is a measure to evaluate how well the nodes are connected together. Depending on the specific applications, many network connectivity measures have been proposed in the past. Examples include the size of a giant connected component (GCC), graph diameter, the mixing time [49], the vulnerability measure [2], and the clustering coefficient [132], each of which often has its own, different mathematical definitions. At the microlevel, network connectivity measures the capacity of edges, paths, loops, some complex motifs [83], or even the centrality of the nodes. Some well-known examples from this category include the epidemic threshold [13], the natural connectivity (i.e., the robustness) [51], degree centrality [39], etc.

In this book, we propose a unified framework to evaluate the connectivity in both single-layered networks and multi-layered networks.

2.1 SINGLE-LAYERED NETWORK MEASURES

In this section, we introduce the connectivity measure framework in single-layered networks, which are called SUBLINE connectivity measures.

SubLine Connectivity Measures
The key to SUBLINE is to view the connectivity of the entire network (G) as the aggregation over the connectivity measures of its sub-networks (e.g., subgraphs), that is

$$C(G, f) = \sum_{\pi \subseteq G} f(\pi), \tag{2.1}$$

where π is a subgraph of G. The non-negative function $f : \pi \to \mathbb{R}^+$ maps any subgraph in G to a non-negative real number and $f(\phi) = 0$ for empty set ϕ. In other words, we view the connectivity of the entire network ($C(G, f)$) as the sum of the connectivity of all the valid subgraphs ($f(\pi)$). The definition in Eq. (5.1) can be used to measure the connectivity of the *entire* network. It can be further extended to measure the local connectivity of a subset of nodes

---

[1]In this book, "graph" and "network" are interchangeably equivalent.

$\mathcal{T}$, where we define $f(\pi) > 0$ iff π is incident to the node set $\mathcal{T}$, i.e.,

$$C_{\mathcal{T}}(G, f) = \sum_{\pi \cap \mathcal{T} \neq \phi} f(\pi). \tag{2.2}$$

It is worth mentioning that motifs (defined in [83]) are subgraphs as well. Here we give three prominent examples for SubLine connectivity: (1) the path capacity, (2) the triangle capacity, and (3) the loop capacity.

Path Capacity. A natural way to measure network connectivity is through path capacity, which measures the total number of (weighted) paths in the network. In this case, the corresponding function $f()$ can be defined as follows:

$$f(\pi) = \begin{cases} \beta^{len(\pi)} & \text{if } \pi \text{ is a valid path of length } len(\pi) \\ 0 & \text{otherwise,} \end{cases} \tag{2.3}$$

where β is a damping factor between $(0, 1/\lambda_G)$ to penalize longer paths. With such an $f()$ function, the connectivity function $C(G, f)$ defined in Eq. (5.1) can be written as

$$C(G, f) = \mathbf{1}' \left(\sum_{t=1}^{\infty} \beta^t \mathbf{A}^t \right) \mathbf{1} = \mathbf{1}'(\mathbf{I} - \beta \mathbf{A})^{-1} \mathbf{1}. \tag{2.4}$$

Remarks. We can also define the path capacity with respect to a given path length t as $C(G, f) = \mathbf{1}'\mathbf{A}^t\mathbf{1}$. When $t = 1$, $C(G, f)$ is reduced to the edge capacity (density) of the graph, it is an important metric for network analysis. On the other hand, the "average" path capacity $(\mathbf{1}'\mathbf{A}^t\mathbf{1})^{1/t}$ of a network converges to the leading eigenvalue of its adjacency matrix, i.e., $(\mathbf{1}'\mathbf{A}^t\mathbf{1})^{1/t} \xrightarrow{t \to \infty} \lambda_G$. It is worth noting that the eigenvalues of a graph's adjacency matrix is used to measure the *Epidemic Threshold* of the contact network [44] in epidemiology. Specifically, the epidemic threshold of the network is defined as the inverse of its leading eigenvalue (i.e., $1/\lambda_G$). Networks with larger leading eigenvalue λ_G (i.e., smaller epidemic threshold) imply that infectious diseases more easily to become a pandemic over the whole population. Considering the two examples in Figure 2.1, the star-shaped network is more vulnerable to epidemic diseases as its average path length is shorter than the link-shaped network. Correspondingly, it λ_G is larger than the one in the link-shaped network.

Triangle Capacity. The number of triangles in a graph plays an important role in calculating *clustering coefficient* and related attributes. By the definition of SubLine connectivity, triangle capacity can be modeled by setting the function $f()$ as

$$f(\pi) = \begin{cases} 1 & \text{if } \pi \text{ is a triangle} \\ 0 & \text{otherwise.} \end{cases} \tag{2.5}$$

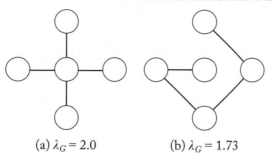

(a) $\lambda_G = 2.0$ (b) $\lambda_G = 1.73$

Figure 2.1: Illustrative examples of epidemic threshold $(1/\lambda_G)$.

The brute-force algorithm for counting the number of triangles in a network is of complexity $O(n^3)$, where n is the number of nodes in the network. The state-of-the-art algorithm has reduced the complexity to $O(n^{2.373})$ [134], but this is still not a scalable algorithm on real-world large datasets. In [125], Tsourakakis showed that the number of triangles in a graph($\triangle(G)$) can be calculated using Eq. (2.6):

$$\triangle(G) = \frac{1}{6} trace\left(\mathbf{A}^3\right) = \frac{1}{6}\sum_{i=1}^{n}\lambda_i^3,\tag{2.6}$$

where $\mathbf{A}$ is the adjacency matrix of the network and λ_i is the top ith eigenvalue (by absolute value) of $\mathbf{A}$. By Eq. (2.6), the triangle capacity $\triangle(G)$ therefore becomes a function of eigenvalues. In real-world graphs, usually, we only need top k eigenvalues to achieve a good approximation for triangle counting. For example, experiments in [125] showed that picking the top 30 eigenpairs can achieve an accuracy of at least 95% in most graphs.

Loop Capacity. Another important way to measure network connectivity is through loop capacity, which measures the total number of (weighted) loops in the network. In this case, the corresponding function $f()$ can be defined as follows:

$$f(\pi) = \begin{cases} 1/len(\pi)! & \text{if } \pi \text{ is a valid loop of length } len(\pi) \\ 0 & \text{otherwise.} \end{cases}\tag{2.7}$$

Then, the connectivity function $C(G, f)$ can be written as

$$C(G, f) = \sum_{t=1}^{\infty}\frac{1}{t!}trace(\mathbf{A}^t) = \sum_{i=1}^{n}e^{\lambda_i}.\tag{2.8}$$

It is interesting to find that the loop capacity defined above is equivalent to the *Natural Connectivity* introduced in [135]. Specifically, the natural connectivity of network G ($S(G)$) is

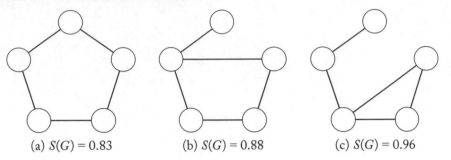

(a) $S(G) = 0.83$ (b) $S(G) = 0.88$ (c) $S(G) = 0.96$

Figure 2.2: Illustrative examples of natural connectivity.

defined as

$$S(G) = \ln \left(\frac{1}{n} \sum_{i=1}^{n} e^{\lambda_i} \right), \tag{2.9}$$

which is commonly used to measure the network tolerance under random failures and external attacks. Empirically, in [14], Chan et al. showed that top k ($k = 50$ in their study) eigenvalues are sufficient for estimating the natural connectivity score. Figure 2.2 shows the illustrative examples of *Natural Connectivity*. Particularly, the natural connectivity ($S(G)$) of the network increases as the length of the loop decreases. This is due to the fact that longer loops are more vulnerable to failures or attacks, as the breakdown of any involved nodes would destroy the entire loop structure.

SubLine Connectivity under Different Network Models. To study how SubLine connectivity changes under different network models, we compute the leading eigenvalue (λ_G), triangle capacity ($\triangle(G)$), and natural connectivity ($S(G)$) in networks that are constructed by the random graph model [35], the small-world network model [133] and preferential attachment network model [4], respectively. To make a fair comparison between models, we generated networks of 1,000 nodes for each model and tuned the corresponding parameters to make sure that the average degree in each network is approximately 30. The results are as shown in Table 2.1, which is averaged over 10 runs. It is worth noting that the leading eigenvalue (λ_G) and the natural connectivity ($S(G)$) in the networks under the preferential attachment model are higher than the ones under the other two models. This is due to the fact that most of the nodes in preferential attachment networks are clustered around several high degree nodes, which significantly shorten the loop sizes and the average distance between node pairs. The triangle capacity, on the other hand, peaks in the small-world networks. In small-world networks, the nodes are of the same degree (which is 30 in our simulations), and each of them is directly connected to 30 closest neighbors. This type of construction scheme would form a great number of closed triads among the neighborhood, thus remarkably increasing the triangle capacity in the network.

Table 2.1: SᴜʙLɪɴᴇ connectivity under different network models

	Random Graph Model	Small World Model	Preferential Attachment Model
λ_G	30.88	30.0	49.94
$\Delta(G)$	4431.8	105,000.0	17,340.4
$S(G)$	23.96	25.18	41.64

Eigen-Function Based Connectivity Measures

The three examples of SᴜʙLɪɴᴇ connectivity suggest that many of the connectivity measures can be estimated through the eigenvalues of the underlying network, which leads to a special group of measures called *Eigen-function based connectivity measures*. Formally, let Λ, U be the eigenvalue and eigenvector matrices of network G, then the eigen-function-based connectivity of G can be defined as

$$C(G, g) = g(\Lambda, U),\tag{2.10}$$

where $g : (\Lambda, U) \rightarrow \mathbb{R}^+$ is a function that maps the eigen-pair of the network to a non-negative connectivity attribute or attribute vector.

Important Eigen-Functions:

In addition to the *Epidemic Threshold* (function of leading eigenvalue), *Triangle Capacity* and *Natural Connectivity* (function of top k leading eigenvalues) mentioned in the previous section, here are some other commonly used eigen-function-based connectivity measures.

Eigenvectors. One of the simplest eigen-function is the eigenvectors of the network:

$$g(\Lambda, U) = U.\tag{2.11}$$

The eigenvectors can be used to evaluate the centrality of nodes [90] or to detect interesting subgraphs [98].

Eigen-Gap. The eigen-gap of a graph is an important parameter in expander graph theory and is defined as the difference between the largest and second-largest (in module) eigenvalues of the graph (as shown in Eq. (2.12)):

$$g(\Lambda, U) = Gap(G) = \lambda_1 - \lambda_2.\tag{2.12}$$

In the expander graph theory, a graph is considered to have a good expansion property if it is both sparse and highly connected [46]. By Cheeger inequality, the expansion property of a graph is strongly correlated to its eigen-gap [27]. As a result, the eigen-gap of the graph can be used as a measurement for its robustness.

2.2   MULTI-LAYERED NETWORK MEASURES

Multi-layered networks have attracted a lot of research attention in recent years. Different models have been proposed to formulate the multi-layered network data structure. In [31], multi-layered networks are represented as a high-order tensor, which is coupled by a second-order within-layer networks tensor and a second-order cross-layer dependency tensor. While in [106], the corresponding data structure is represented as a quadruplet $M = \{V_M, E_M, V, \mathcal{L}\}$, in which each distinct nodes in V can appear in multiple elementary layers in $\mathcal{L} = \{L_1, \ldots, L_d\}$. Then, $V_M \subseteq V \times L_1 \times \ldots \times L_d$ represents the nodes in each layer, and $E_M = V_M \times V_M$ represents both within-layer and cross-layer links in the entire system. In [9], the model is simplified into a pair $M = (\mathcal{G}, \mathcal{C})$, where $\mathcal{G}$ gives all the within-layer networks and $\mathcal{C}$ provides all the cross-layer dependencies. In [55], Kivela et al. presented a comprehensive survey on different types of multi-layered networks, which include multi-modal networks [45], multi-dimensional networks [7], multiplex networks [5], and interdependent networks [12]. The problem addressed in our study is most related to the interdependent networks. In [101] and [40], the authors presented an in-depth introduction to the fundamental concepts of interdependent multi-layered networks as well as the key research challenges. In a multi-layered network, the failure of a small number of nodes might lead to catastrophic damages to the entire system as shown in [12] and [127]. In [12, 41, 96, 108, 110], different types of *two-layered* interdependent networks were thoroughly analyzed. In [40], Gao et al. analyzed the robustness of multi-layered networks with the star- and loop-shaped dependency structures.

To extend the connectivity measures in single-layered networks to multi-layered networks, we first give the formal definition of multi-layered networks as follows.

Definition 2.1   Multi-layered Network Model (MuLaN). Given (1) a binary $g \times g$ abstract layer-layer dependency network $\mathbf{G}$, where $\mathbf{G}(i, j) = 1$ indicates layer-j depends on layer-i (or layer-i supports layer-j), $\mathbf{G}(i, j) = 0$ means that there is no direct dependency from layer-i to layer-j; (2) a set of within-layer networks $\{G_i\}_{i=1}^g$ with adjacency matrices $\mathcal{A} = \{\mathbf{A}_1, \ldots, \mathbf{A}_g\}$; (3) a set of cross-layer node-node dependency matrices $\mathcal{D}$, indexed by pair (i, j), $i, j \in [1, \ldots, g]$, such that for a pair (i, j), if $\mathbf{G}(i, j) = 1$, then $\mathbf{D}_{i,j}$ is a $n_i \times n_j$ matrix; otherwise $\mathbf{D}_{i,j} = \phi$ (i.e., an empty set); (4) θ is a one-to-one mapping function that maps each node in layer-layer dependency network $\mathbf{G}$ to the corresponding within-layer adjacency matrix $\mathbf{A}_i$ ($i = 1, \ldots, g$); and (5) φ is another one-to-one mapping function that maps each edge in $\mathbf{G}$ to the corresponding cross-layer node-node dependency matrix $\mathbf{D}_{i,j}$. We define a multi-layered network as a quintuple $\Gamma = < \mathbf{G}, \mathcal{A}, \mathcal{D}, \subseteq, \simeq >$.

For simplicity, we restrict the within-layer adjacency matrices $\mathbf{A}_i$ to be simple (i.e., no self loops), symmetric and binary; and the extension to the weighted, asymmetric case is straightforward. In our work, we require the cross-layer dependency network $\mathbf{G}$ to be an unweighted graph with arbitrary dependency structures. Notice that compared with the existing pair-wise two layered models, MuLaN allows a much more flexible and complicated dependency structure

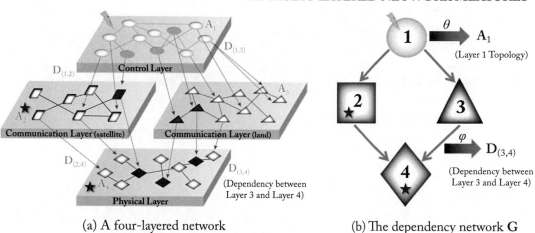

(a) A four-layered network   (b) The dependency network $\mathbf{G}$

Figure 2.3: An illustrative example of MuLaN model.

among different layers. For the cross-layer node-node dependency matrix $\mathbf{D}_{i,j}$, $\mathbf{D}_{i,j}(s,t) = 1$ indicates that node s in layer i supports node t in layer j.

Figure 2.3a presents an example of a four-layered network. In this example, layer-1 (i.e., the control layer) is the supporting layer (i.e., the root node in the layer-layer dependency network $\mathbf{G}$). Layer-2 and layer-3 directly depend on layer-1 (i.e., one represents a communication layer by satellites and the other represents another communication layer in landlines, respectively), while layer-4 (i.e., the physical layer) depends on both communication layers (layer-2 and layer-3). The abstracted layer-layer dependency network ($\mathbf{G}$) is shown in Figure 2.3b. $\mathcal{A} = \{\mathbf{A}_1, \mathbf{A}_2, \mathbf{A}_3, \mathbf{A}_4\}$ denotes the within-layer adjacency matrices, each of which describes the network topology in the corresponding layer. In this example, $\mathcal{D}$ is a set of matrices containing only four non-empty matrices: $\mathbf{D}_{(1,2)}$, $\mathbf{D}_{(1,3)}$, $\mathbf{D}_{(2,4)}$, and $\mathbf{D}_{(3,4)}$. For example, $\mathbf{D}_{(3,4)}$ describes the node-node dependency between layer-3 and layer-4. The one-to-one mapping function θ maps node 1 in $\mathbf{G}$ (i.e., layer-1) to the within-layer adjacency matrix of layer-1 ($\mathbf{A}_1$); and the one-to-one mapping function φ maps edge $\langle 3, 4 \rangle$ in $\mathbf{G}$ to the cross-layer node-node dependency matrix $\mathbf{D}_{(3,4)}$ as shown in Figure 2.3b. Different from existing multi-layered network models, the formulation used in our study gives more emphasis on the abstracted dependency network structure $\mathbf{G}$, which makes it easier to unravel the impact path for a set of nodes from a given layer.

With the above definition, the connectivity of a multi-layered network Γ can be defined as a weighted summation over the connectivity scores from all the layers

$$C(\Gamma) = \sum_{i=1}^{g} \alpha_i C(G_i, f_i),$$

(2.13)

where α_i is the weight for layer i.

CHAPTER 3

Connectivity Inference Computation

In real-world applications, networks are evolving over time. Moreover, the input network structure is often incomplete due to noise and accessibility issues. The research on network connectivity inference can be categorized into two parts: (1) network property tracking in single-layered networks and (2) inference in multi-layered networks.

Various network properties have been studied under dynamic settings. In [66], Leskovec et al. discovered the growth pattern of network density and diameter in real networks. In [122], two online algorithms were provided for tracking node proximity and centrality on bipartite graphs. In [73], a graph kernel tracking algorithm was proposed for dynamic networks. The other area of research that is related to our work is evolutionary spectral clustering on graphs. In [93], Ning et al. proposed an incremental spectral clustering algorithm based on an iterative updates on the eigen-system of the network. In [17], Chen et al. proposed an efficient online algorithm to track some important network connectivity measures (e.g., the leading eigenvalue, the robustness measure) on a temporal dynamic network. For the multi-layered network inference problem, a collaborative filtering-based method is proposed to infer the missing cross-layer dependencies in [20]. Other remotely related studies include cross-network ranking [92] in multi-layered networks and multi-view data analysis [71, 136, 144].

In this book, we propose to study the following two problems: (1) eigen-functio-based connectivity measures tracking in dynamic networks and (2) cross-layer dependence inference in multi-layered networks.

3.1 EIGEN-FUNCTIONS TRACKING IN DYNAMIC NETWORKS

As we have shown in Chapter 2, many node centrality measures and critical network connectivity measures can be well approximated by some well-defined eigen-functions. For example, in node centrality analysis, one commonly used parameter is eigenvector centrality [90], which is defined with the leading eigenvector of the graph. As for network connectivity, frequently used parameters include epidemic threshold [13, 97, 131], clustering coefficient [132], natural connectivity [2, 14, 37], eigen-gap, etc.

Most of the graph parameters mentioned above are all based on static graphs. However, in real-world applications, the graph structure evolves over time. In some cases, subtle changes in the graph structure may lead to huge differences in some of its properties. For example, when the Ebola virus was first brought to the U.S. continent, some emerging connections in the contact network would greatly reduce the epidemic threshold of the graph, and eventually cause the outbreak of the disease. By monitoring those key parameters as the graph evolves and analyzing the attribution for sharp parameter changes timely, we would be able to get prepared for emergent events at an early stage. Another application scenario is the social network. In websites like Facebook and Twitter, new connections between users emerge all the time, which would, in turn, change the influential individuals in the network. It is crucial for online marketing companies to keep track of those changes since their ad-targeting strategies may need to be modified accordingly.

For the eigen-functions tracking problem, simply re-computing the eigen-pairs whenever the graph structure changes is computationally costly over fast-changing large graphs. The popular Lanczos method for computing top-k eigen-pairs would require $O(mk + nk^2)$ time, where m and n are the numbers of edges and nodes in the graph. Although the complexity seems acceptable for one-time calculation in static graphs, it would be too expensive for large dynamic graphs. To address this challenge, we consider a way of updating the eigen-pairs incrementally instead of re-computing them from scratch at each time stamp. In this work, we propose two online algorithms to track the eigen-pairs of a dynamic graph efficiently, which bear linear time complexities w.r.t. the number of nodes n in the graph and the number of changed edges s at the current stamp. Based on these algorithms, we introduce a general attribution analysis framework for identifying key connection changes that have the largest impact on the graph. Last, to control the accumulated tracking error of eigen-functions, we propose an error estimation method to detect sharp error increase timely so that the accumulated error can be eliminated by restarting the tracking algorithms.

In addition to the problem definition, the main contributions of this work can be summarized as follows.

- *Algorithms*. We propose two online algorithms to track the top eigen-pairs of a dynamic graph, which in turn enable us to track a variety of important network parameters based on certain eigen-functions. In addition, we provide a framework for attribution analysis of eigen-functions and a method for estimating tracking errors.

- *Evaluations*. We evaluate our methods with other eigen-pair update algorithms on real-world datasets to validate the effectiveness and efficiency of the proposed algorithms.

3.1.1   PROBLEM DEFINITION

In this section, we introduce the notations, followed by a formal definition of the eigen-function tracking problem.

Table 3.1: Symbols used in TRIP-BASIC and TRIP

Symbol	Definition and Description
$G^t(V, E)$	undirected, unipartite network at time t
m	number of edges in the network
n	number of nodes in the network
$\mathbf{B}, \mathbf{C}, \ldots$	matrices (bold upper case)
$\mathbf{b}, \mathbf{c}, \ldots$	vectors (bold lower case)
$\mathbf{A}^t$	adjacency matrix of $G^t(V, E)$ at time t
$\Delta \mathbf{A}^t$	perturbation matrix from time t to $t+1$
$\Delta(G^t)$	number of triangles in G^t
$S(G^t)$	robustness score of G^t
$Gap(G^t)$	eigen-gap of G^t
$(\lambda_j^t, \mathbf{u_j}^t)$	jth eigen-pair of $\mathbf{A}^t$
$[\Delta \mathbf{A}^t]_{t=t_1 \ldots t_2}$	perturbation matrices of dynamic graph from time t_1 to t_2
$[(\Lambda_k^t, \mathbf{U}_k^t)]_{t=t_1 \ldots t_2}$	top k eigen-pairs from time t_1 to t_2
$[\Delta(G^t)]_{t=t_1 \ldots t_2}$	$\Delta(G)$ from time t_1 to t_2
$[S(G^t)]_{t=t_1 \ldots t_2}$	$S(G)$ from time t_1 to t_2

The symbols used in this work is shown in Table 3.1. We consider the graph in each time stamp $G^t(V, E)$ is undirected and unipartite. Consistent with standard notation, we use bold uppercase for matrices (e.g., $\mathbf{B}$), and bold lowercase for vectors (e.g., $\mathbf{b}$). For each time stamp, the graph is represented by its adjacency matrix $\mathbf{A}^t$. $\Delta \mathbf{A}^t$ denotes the perturbation matrix from time t to $t + 1$. $(\lambda_j^t, \mathbf{u_j^t})$ is the jth eigen-pair of $\mathbf{A}^t$. The number of triangles and robustness score of the graph at time t are represented as $\Delta(G^t)$ and $S(G^t)$, respectively.

With the above notations, the eigen-function is defined as a function that maps eigen-pairs of the graph to certain graph attribute or attribute vector, which can be expressed as

$$g : (\Lambda_k, \mathbf{U}_k) \rightarrow \mathbb{R}^x (x \in \mathbb{N}). \tag{3.1}$$

The simplest eigen-function is the eigenvalues and eigenvectors themselves. Specifically, the eigenvalues of a graph's adjacency matrix can be used to measure the epidemic threshold or path capacity of a graph as mentioned in Chapter 2, while the eigenvectors can be used to evaluate the centrality of nodes [90] or to detect interesting subgraphs [98]. Some more sophisticated eigen-functions include the number of triangles, network robustness (i.e., natural connectivity), eigen-gap, etc.

For all the above-mentioned network parameters (e.g., epidemic threshold, eigenvector centrality, number of triangles, robustness measurement, eigen-gap), it is often sufficient to use top-k eigen-pairs to achieve a high accuracy estimation of these parameters. Therefore, in order to track these parameters on a dynamic graph, we only need to track the corresponding top-k eigen-pairs at each time stamp. Formally, the eigen-function tracking problem is defined as follows. Once the top-k eigen-pairs are estimated, we can use Eqs. (2.11)–(2.12) to update the corresponding eigen-functions.

Problem 3.1   Top-k Eigen-Pairs Tracking

Given: (1) A dynamic graph G tracked from time t_1 to t_2 with starting matrix $\mathbf{A}^{t_1}$, (2) an integer k, and (3) a series of perturbation matrices $[\Delta\mathbf{A}^t]_{t=t_1,\dots t_2-1}$.

Output: The corresponding top-k eigen-pairs at each time stamp $[(\Lambda_k^t, \mathbf{U}_k^t)]_{t=t_1,\dots,t_2}$.

3.1.2   PROPOSED ALGORITHMS

In this section, we present our solutions for Problem 3.1. We start with a baseline solution (TRIP-BASIC) and then present its high-order variant (TRIP), followed by the attribution analysis framework for different eigen-functions and an error estimation method.

Key Idea

The key idea for TRIP-BASIC and TRIP is to incrementally update the eigen-pairs with corresponding perturbation terms at each time stamp. By matrix perturbation theory [114], we have the following perturbation equation:

$$(\mathbf{A}^t + \Delta\mathbf{A}^t)(\mathbf{u}_j^t + \Delta\mathbf{u_j}) = (\lambda_j^t + \Delta\lambda_j)(\mathbf{u}_j^t + \Delta\mathbf{u_j}). \tag{3.2}$$

As the perturbation matrix is often very sparse, it is natural to assume that graphs in two consecutive time stamps share a fixed eigen-space. Therefore, the perturbation eigenvector $\Delta\mathbf{u_j}$ can be expressed as $\Delta\mathbf{u_j} = \sum_{i=1}^{k} \alpha_{ij}\mathbf{u}_i^t$, which is the linear combination of old eigenvectors. Taking the two-dimensional eigenspace in Figure 3.1 as an example, the old eigenvectors are $\mathbf{u}_1^t$ and $\mathbf{u}_2^t$ marked in orange; the new eigenvectors $\mathbf{u}_1^{t+1}$ and $\mathbf{u}_2^{t+1}$ (in green) can be decomposed into old eigenvectors $\mathbf{u}_1^t$, $\mathbf{u}_2^t$ and perturbation eigenvectors $\Delta\mathbf{u}_1$, $\Delta\mathbf{u}_2$ in the same plane. Expanding Eq. (3.2), we get

$$\mathbf{A}^t\mathbf{u}_j^t + \Delta\mathbf{A}^t\mathbf{u}_j^t + \mathbf{A}^t\Delta\mathbf{u_j} + \Delta\mathbf{A}^t\Delta\mathbf{u_j}$$
$$= \lambda_j^t\mathbf{u}_j^t + \Delta\lambda_j\mathbf{u}_j^t + \lambda_j^t\Delta\mathbf{u_j} + \Delta\lambda_j\Delta\mathbf{u_j}.$$

By the fact that $\mathbf{A}^t\mathbf{u}_j^t = \lambda_j^t\mathbf{u}_j^t$, the perturbation equation can be simplified as

$$\Delta\mathbf{A}^t\mathbf{u}_j^t + \mathbf{A}^t\Delta\mathbf{u_j} + \Delta\mathbf{A}^t\Delta\mathbf{u_j}$$
$$= \Delta\lambda_j\mathbf{u}_j^t + \lambda_j^t\Delta\mathbf{u_j} + \Delta\lambda_j\Delta\mathbf{u_j}. \tag{3.}$$

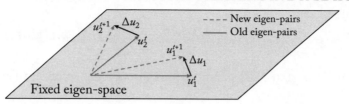

Figure 3.1: Incremental update for eigen-pairs tracking.

Multiplying the term $\mathbf{u}_j^{t\,\prime}$ on both sides, as eigenvectors are of unit length, we have

$$\mathbf{u}_j^{t\,\prime}\Delta\mathbf{A}^t\mathbf{u}_j^t + \mathbf{u}_j^{t\,\prime}\Delta\mathbf{A}^t\Delta\mathbf{u}_j = \Delta\lambda_j + \mathbf{u}_j^{t\,\prime}\Delta\lambda_j\Delta\mathbf{u}_j. \tag{3.4}$$

As we assume that $\Delta\mathbf{u}_j \ll \mathbf{u}_j$ and $\Delta\lambda_j \ll \lambda_j$, the high-order terms $\mathbf{u}_j^{t\,\prime}\Delta\mathbf{A}^t\Delta\mathbf{u}_j$ and $\mathbf{u}_j^{t\,\prime}\Delta\lambda_j\Delta\mathbf{u}_j$ in Eq. (3.4) can be discarded without losing too much accuracy. Therefore, $\Delta\lambda_j$ can be estimated as

$$\Delta\lambda_j = \mathbf{u}_j^{t\,\prime}\Delta\mathbf{A}^t\mathbf{u}_j^t. \tag{3.5}$$

The difference between Trip-Basic and Trip lies in their ways of estimating perturbation eigenvectors, which will be discussed below.

Trip-Basic

The Trip-Basic algorithm is a first-order eigen-pair tracking method, which ignores the high-order terms in the perturbation equation when updating eigenvectors at each time stamp. By removing the high-order terms, the perturbation equation Eq. (3.3) can be written as

$$\Delta\mathbf{A}^t\mathbf{u}_j^t + \mathbf{A}^t\Delta\mathbf{u}_j = \Delta\lambda_j\mathbf{u}_j^t + \lambda_j^t\Delta\mathbf{u}_j.$$

Replacing all $\Delta\mathbf{u}_j$ terms with $\sum_{i=1}^k \alpha_{ij}\mathbf{u}_i^t$ and multiplying the term $\mathbf{u}_p^{t\,\prime}$ $(p \neq j)$ on both sides, by applying the orthogonality property of eigenvectors to the new equation, we can solve the coefficient α_{pj} as

$$\alpha_{pj} = \frac{\mathbf{u}_p^{t\,\prime}\Delta\mathbf{A}^t\mathbf{u}_j^t}{\lambda_j^t - \lambda_p^t}.$$

Therefore, $\Delta\mathbf{u}_j$ can be estimated as

$$\Delta\mathbf{u}_j = \sum_{i=1, i\neq j}^k \left(\frac{\mathbf{u}_i^{t\,\prime}\Delta\mathbf{A}^t\mathbf{u}_j^t}{\lambda_j^t - \lambda_i^t}\mathbf{u}_i^t \right). \tag{3.6}$$

Suppose $\mathbf{A}^t$ is perturbed with a set of edges $\Delta E = \langle p_1, r_1 \rangle, \ldots, \langle p_s, r_s \rangle$ where s is the number of non-zero elements in perturbation matrix $\Delta\mathbf{A}$. In Eq. (3.6), the term $\mathbf{u}_j^{t\,\prime}\Delta\mathbf{A}\mathbf{u}_j^t$ can be expanded as

$$\mathbf{u}_j^{t'} \Delta \mathbf{A}^t \mathbf{u}_j^t = \sum_{\langle p,r \rangle \in \Delta E} \Delta \mathbf{A}^t (p,r) \mathbf{u}_{pj}^t \mathbf{u}_{rj}^t. \tag{3.7}$$

Equations (3.6) and (3.7) naturally lead to our base solution (TRIP-BASIC) for solving Problem 3.1 as follows.

Algorithm 3.1 TRIP-BASIC: First-Order Eigen-Pairs Tracking

Input: Dynamic graph G tracked from time t_1 to t_2, with starting eigen-pairs $(\Lambda_k^{t_1}, \mathbf{U}_k^{t_1})$, series of perturbation matrices $[\Delta \mathbf{A}^t]_{t=t_1,\dots t_2-1}$

Output: Corresponding eigen-pairs $[(\Lambda_k^t, \mathbf{U}_k^t)]_{t=t_1+1,\dots t_2}$

1: **for** $t = t_1$ to $t_2 - 1$ **do**
2:    **for** $j = 1$ to k **do**
3:       Initialize $\Delta \mathbf{u}_j \leftarrow \mathbf{0}$
4:       **for** $i = 1$ to $k, i \neq j$ **do**
5:          $\Delta \mathbf{u}_j \leftarrow \Delta \mathbf{u}_j + \frac{\mathbf{u}_i^{t'} \Delta \mathbf{A}^t \mathbf{u}_j^t}{\lambda_j^t - \lambda_i^t} \mathbf{u}_i^t$
6:       **end for**
7:       Calculate $\Delta \lambda_j \leftarrow \mathbf{u}_j^{t'} \Delta \mathbf{A}^t \mathbf{u}_j^t$
8:       Update $\lambda_j^{t+1} \leftarrow \lambda_j^t + \Delta \lambda_j$
9:       Update $\mathbf{u}_j^{t+1} \leftarrow \mathbf{u}_j^t + \Delta \mathbf{u}_j$
10:    **end for**
11: **end for**
12: Return $[(\Lambda_k^t, \mathbf{U}_k^t)]_{t=t_1+1\dots t_2}$

The approximated eigen-pairs for each time stamp is computed from steps 2–10. Each $\Delta \lambda_j$ and $\Delta \mathbf{u}_j$ is calculated from steps 3–7 by Eqs. (3.6) and (3.7). At steps 8 and 9, λ_j^t and $\mathbf{u}_j^t$ is updated with $\Delta \lambda_j$ and $\Delta \mathbf{u}_j$. Note that after updating the eigenvector in step 9, we normalize each of them to unit length.

Complexity Analysis. The efficiency of proposed Algorithm 3.1 is summarized in Lemma 3.2. Both time complexity and space complexity are linear w.r.t. the total number of the nodes in the graph (n) and the total number of the time stamps (T).

Lemma 3.2   *Complexity of First-Order Eigen-Function Tracking. Suppose T is the total number of the time stamps, s is the average number of perturbed edges in $[\Delta \mathbf{A}^t]_{t=t_1,\dots t_2-1}$, then the time cost for Algorithm 3.1 is $O(Tk^2(s+n))$; the space cost is $O(Tnk + s)$.*

Proof. In each time stamp from time t_1 to $t_2 - 1$, top k eigen-pairs are updated in steps 2–10. By Eq. (3.7), the complexity of computing term $\mathbf{u}_j^{t'} \Delta \mathbf{A}^t \mathbf{u}_j^t$ is $O(s)$, so the overall complexity of step 5 is $O(s + n)$. Therefore, calculating $\Delta \mathbf{u_j}$ from steps 4–6 takes $O(k(s + n))$. In step 7, computing $\Delta \lambda_j$ takes another $O(s)$. Updating λ_j^t and $\mathbf{u}_j^t$ in steps 8 and 9 takes $O(1)$ and $O(n)$. Therefore, updating all top-k eigen-pairs $\mathbf{U}_k^t$ and Λ_k^t takes $O(k^2(s + n))$ and $O(ks)$, respectively. Thus, the overall time complexity for T iterations is $O(Tk^2(s + n))$.

For space cost, it takes $O(k)$ and $O(nk)$ to store Λ_k^t and $\mathbf{U}_k^{t}$ at each time stamp. In the update phase from steps 2–10, it takes $O(s)$ to store $\Delta \mathbf{A}^t$, $O(1)$ to update λ_j^t and $O(n)$ to update $\mathbf{u}_j^t$. However, the space used in the update phase can be reused in each iteration. Therefore, the overall space complexity for T time stamps takes a space of $O(Tnk + s)$. $\square$

Trip

The baseline solution in Algorithm 3.1 is simple and straightforward, but it has the following limitations. First, the approximation error of first-order matrix perturbation is in the order of $\|\Delta \mathbf{A}^t\|$. In other words, the quality of such approximation might decrease quickly w.r.t. the increase of $\|\Delta \mathbf{A}^t\|$. Second, the approximation quality is highly sensitive to the small eigen-gap of $\mathbf{A}^t$ as indicated by Eq. (3.6). In order to address these limitations, we further propose Algorithm 3.2 by adopting the high-order matrix perturbation to update the eigen-pairs of $\mathbf{A}^{t+1}$. The main difference between Algorithms 3.2 and 3.1 is that we take high-order terms in the perturbation equation (Eq. (3.3)) into consideration while updating eigenvectors. Similar to TRIP-BASIC we replace all $\Delta \mathbf{u_j}$ terms with $\sum_{i=1}^{k} \alpha_{ij} \mathbf{u_i^t}$ and multiplying the term $\mathbf{u_p^t}'$ (for $1 \leq p \leq k$, $p \neq j$) on both sides. By applying the orthogonality property of eigenvectors to the new equation, we have

$$\mathbf{X}^t(p, j) + \alpha_{pj} \lambda_p^t + \sum_{i=1}^{k} \mathbf{X}^t(p, i)\alpha_{ij} = \alpha_{pj} \lambda_j^t + \alpha_{pj} \Delta \lambda_j,$$

where $\mathbf{X}^t = \mathbf{U}_k^t{}' \Delta \mathbf{A}^t \mathbf{U}_k^t$. Reorganizing the terms in the above equation, we have

$$\mathbf{X}^t(p, j) - \alpha_{pj}(\lambda_j^t + \Delta \lambda_j - \lambda_p^t) + \sum_{i=1}^{k} \mathbf{X}^t(p, i)\alpha_{ij} = 0.$$

By defining $\mathbf{v} = \lambda_j^t + \Delta \lambda_j - \lambda_p^t$ for $p = 1, \ldots, k$, $\mathbf{D}^t = diag(\mathbf{v})$ and $\boldsymbol{\alpha}_j = [\alpha_{1j}, \ldots, \alpha_{kj}]$, the above equation can be expressed as

$$\mathbf{X}^t(:, j) - \mathbf{D}^t \boldsymbol{\alpha}_j + \mathbf{X}^t \boldsymbol{\alpha}_j = 0.$$

Solve the above equation for $\boldsymbol{\alpha}_j$, we have

$$\boldsymbol{\alpha}_j = (\mathbf{D}^t - \mathbf{X}^t)^{-1} \mathbf{X}^t(:, j).$$

---

[1]Here the *diag* function works the same as the one in Matlab. When applying to a matrix, *diag* returns a vector of the main diagonal elements of the matrix; when applying to a vector, it returns a square diagonal matrix with the elements of the vector on the main diagonal.

Algorithm 3.2 Trip: High-Order Eigen-Pairs Tracking

Input: Dynamic graph G tracked from time t_1 to t_2, with starting eigen-pairs $(\Lambda_k^{t_1}, U_k^{t_1})$, series
   of perturbation matrices $[\Delta A^t]_{t=t_1,\ldots t_2-1}$
Output: Corresponding eigen-pairs $[(\Lambda_k^t, U_k^t)]_{t=t_1+1,\ldots t_2}$

1: **for** $t = t_1$ to $t_2 - 1$ **do**
2:     Calculate $X^t \leftarrow U_k^{t'} \Delta A^t U_k^t$
3:     $\Delta \Lambda_k \leftarrow diag(X^t)^1$
4:     Update $\Lambda_k^{t+1} \leftarrow \Lambda_k^t + \Delta \Lambda_k$
5:     **for** $j = 1$ to k **do**
6:         Calculate $v \leftarrow \lambda_j^t + \Delta \lambda_j - \lambda_p^t$ for $p = 1, \ldots, k$
7:         $D^t \leftarrow diag(v)$
8:         Calculate $\alpha_j \leftarrow (D^t - X^t)^{-1} X(:, j)$
9:         Calculate $\Delta u_j \leftarrow \sum_{i=1}^k \alpha_{ij} u_i^t$
10:        Update $u_j^{t+1} \leftarrow u_j^t + \Delta u_j$
11:    **end for**
12: **end for**
13: Return $[(\Lambda_k^t, U_k^t)]_{t=t_1+1\ldots t_2}$

In Algorithm 3.2, the top-k eigen-pairs at each time stamp is updated from steps 2–11. In step 2, matrix X^t is calculated for computing $\Delta \Lambda_k$ and ΔU_k. In step 4, all top-k eigenvalues Λ_k are updated by $\Delta \Lambda_k$. From steps 6–10, each u_j^t is updated according to the derivations of the eigen-update rule mentioned above. Again, after we update the eigenvector in step 9, we normalize each of them to unit length.

Complexity Analysis. The efficiency of Algorithm 3.2 is given in Lemma 3.3. Compared with Trip-Basic, both time and space complexity are still linear w.r.t. the total number of nodes in the graph and the total number of time stamps, with a slight increase in k, which is often very small.

Lemma 3.3 *Complexity of High-Order Eigen-Function Tracking. Suppose T is the total number of time stamps, s is the average number of perturbed edges in $[\Delta A^t]_{t=t_1,\ldots t_2-1}$, then the time cost for Algorithm 3.2 is $O(T(k^4 + k^2(n + s)))$; the space cost is $O(Tnk + k^2 + s)$.*

Proof. In each time stamp from time t_1 to $t_2 - 1$, top k eigen-pairs are updated in steps 2–11. Using the update rule provided in Eq. (3.7), calculating X^t in step 2 takes $O(k^2s)$. Updating to eigenvalues in steps 3–4 takes $O(k)$. From steps 5–11, eigenvectors are updated. It takes $O(k^3)$ in to do matrix inversion and multiplication in step 8 and $O(nk)$ to calculate Δu_j in step 9. Therefore, updating U_k^t takes $O(k^4 + nk^2))$. Thus, the overall time complexity for T iterations takes $O(T(k^4 + k^2(n + s)))$.

For space cost, it takes $O(k)$ and $O(nk)$ to store Λ_k^t and U_k^t, $O(s)$ to store ΔA^t for each time stamp. In the update phase from steps 2–11, it takes $O(k^2)$ to store and calculate X^t, D^t; $O(k)$ to store v and α_j; $O(k^2)$ to calculate α_j. However, the space cost in the update phase can be reused in each iteration. Therefore, the overall space complexity for T time stamps takes a space of $O(Tnk + k^2 + s)$. $\qquad\square$

Attribution Analysis

Based on our TRIP algorithms, we can effectively track the corresponding eigen-functions of interest. In reality, we might also be interested in understanding the key factors that cause these changes in dynamic graphs. For example, among all the changed edges in ΔA, which edge is most important in causing the increase/decrease of the epidemic threshold, or the number of triangles, etc. The importance of an edge $\langle p, r \rangle \in \Delta E$ can be measured as the change it can make on the corresponding eigen-functions, which can be written as

$$score(\langle p, r \rangle) \sim \Delta g_{\langle p, r \rangle} = g_{G \cup \langle p, r \rangle} - g_G,$$

where $g(.)$ is one of the eigen-functions we defined in Section 3.1.1.

Algorithm 3.3 Dynamic Attribution Analysis

Input: Dynamic graph G and eigen-pairs (Λ_k^t, U_k^t) at time t, perturbation matrix ΔA^t, eigen-function $g(.)$, number l

Output: Top l added edges and removed edges at time t that have the largest impact on eigen-function $g(.)$

1: $removed \leftarrow$ extract all removed edges in ΔA^t
2: $added \leftarrow$ extract all added edges in ΔA^t
3: **for** each edge $\langle p, r \rangle$ in $removed$ **do**
4:     $score(\langle p, r \rangle) \leftarrow g_{G^t} - g_{G^t \setminus \langle p, r \rangle}$
5: **end for**
6: **for** each edge $\langle p, r \rangle$ in $added$ **do**
7:     $score(\langle p, r \rangle) \leftarrow g_{G^t \cup \langle p, r \rangle} - g_{G^t}$
8: **end for**
9: Return top l edges in $removed$ and $added$ with highest scores, respectively.

In Algorithm 3.3, all removed edges and added edges are extracted from ΔA in steps 1 and 2. The impact score of each removed edge at time t is calculated from steps 3–5. Similarly, the score of each added edge is calculated from steps 6–8. In the end, top l removed edges and l added edges are returned as high impact edges at time t.

Complexity Analysis. Assume that the complexity of calculating $\Delta g_{\langle p, r \rangle}$ is $h(n, k, s)$, where h is a function of number of nodes n, number of eigen-pairs k and number of changed edges s.

Then the complexity of calculating the impact scores of all changed edges (from steps 3–8) is $O(sh(n, k, s))$. Given the impact score of each changed edges, the complexity of picking out top l edges from *removed* and *added* set using heap structure is $O(|removed|\log l) + O(|added|\log l) = O(s\log l)$. Therefore, the overall complexity for attribution analysis at time t is $O(s(h(n, k, s) + \log l))$.

Error Estimation

The core mechanism for both TRIP-BASIC and TRIP is to incrementally update the eigen-pairs at each time stamp. With this scheme, the tracking error of eigen-pairs would accumulate as time goes by. Therefore, finding a proper time to restart the algorithm is of key importance to keep the tracking error within a reasonable range. For simplicity, we only estimate the error of the leading eigenvalue since it is the key part for most of the eigen-functions. Here we denote $err(\lambda^t)$ as the estimated error on λ introduced at time t. Intuitively, $err(\lambda^t)$ would be strongly correlated to the impact of $\Delta \mathbf{A}^t$ on the original eigenspace. As the original eigenspace is defined by the top-k eigenvectors $\mathbf{U}_k^{t_1}$ at the first time stamp t_1, to measure the impact of $\Delta \mathbf{A}^t$ on $\mathbf{U}_k^{t_1}$, we can project $\Delta \mathbf{A}^t$ into this space and take the Frobenius norm of the projection as its actual impact. Equation (3.8) formalizes the impact function of $\Delta \mathbf{A}^t$ on eigenspace $\mathbf{U}_k^{t_1}$:

$$err(\lambda^t) \sim impact(\Delta \mathbf{A}^t, \mathbf{U}_k^{t_1}) = \| \mathbf{U}_k^{t_1} \mathbf{U}_k^{t_1'} \Delta \mathbf{A}^t \|_{Fro} . \tag{3.8}$$

We denote the summation of the perturbation impacts from the first time stamp t_1 to current stamp t as $err_{acc}(\lambda^t)$. This number can be viewed as a good approximation of accumulated tracking error on leading eigenvalue from t_1 to t. In other words, the curve of $err_{acc}(\lambda^t)$ from $t = t_1, \ldots, t_2$ would have a similar shape with real tracking error curve of TRIP algorithms.

Algorithm 3.4 Error Estimation for Eigen-function Tracking

Input: Dynamic graph G tracked from time t_1 to t_2, with starting eigen-pairs $(\Lambda_k^{t_1}, \mathbf{U}_k^{t_1})$, series of perturbation matrices $[\Delta \mathbf{A}^t]_{t=t_1, \ldots t_2-1}$

Output: Corresponding estimated error $err_{acc}(\lambda^t)$ for $t = t_1 + 1, \ldots t_2$

1: Initialize $err_{acc}(\lambda^{t_1}) \leftarrow 0$

2: Initialize $P \leftarrow \mathbf{U}_k^{t_1} \mathbf{U}_k^{t_1'}$

3: **for** $t = t_1 + 1$ to t_2 **do**

4:     Calculate $impact(\Delta \mathbf{A}^t, \mathbf{U}_k^{t_1}) \leftarrow \| P\Delta \mathbf{A}^t \|_{Fro}$

5:     Calculate $err_{acc}(\lambda^t) \leftarrow err_{acc}(\lambda^{t-1}) + impact(\Delta \mathbf{A}^t, \mathbf{U}_k^{t_1})$

6: **end for**

7: Return $err_{acc}(\lambda^t)$ for $t = t_1 + 1, \ldots t_2$

In Algorithm 3.4, $err_{acc}(\lambda^{t_1})$ is initialized as 0 in step 1 and P is initialized as the projection matrix in step 2. From steps 3–6, the impact of each perturbation matrix is calculated and accumulated to $err_{acc}(\lambda^t)$. In step 7, the estimated error array $err_{acc}(\lambda^t)$ for $t = t_1 + 1, \ldots t_2$ is returned.

Complexity Analysis. The complexity of initializing projection matrix P is $O(n^2 k)$. Since $\Delta \mathbf{A}^t$ is often very sparse, the complexity of calculating $impact(\Delta \mathbf{A}^t, \mathbf{U}_k^{t_1})$ can be reduced to $O(ns)$ where s is the number of changed edges at current time stamp. The complexity of accumulating $err_{acc}(\lambda^t)$ at each time stamp is $O(1)$. Therefore, the overall time complexity for error estimation over time series of length T is $O(n^2 k + Tns)$.

3.1.3   EXPERIMENTAL EVALUATION

In this section, we evaluate TRIP-BASIC and TRIP on real datasets. All the experiments are designed to answer the following two questions.

- *Effectiveness*: how accurate are our algorithms in tracking eigen-functions, analyzing corresponding attributions, and estimating the tracking errors?

- *Efficiency*: how fast are the tracking algorithms?

Experiment Setup

Machine. We ran our experiment in a machine with two processors Intel Xeon 3.5 GHz with 256 GB of RAM. Our experiment is implemented with Matlab using a single thread.

Datasets. Here we use three real datasets for evaluations.

AS. The first dataset we use for the evaluation is the Autonomous system graph, which is available at http://snap.stanford.edu/data/. The graph has recorded communications between routers on the Internet for a long period of time. Based on the data from http://www.routeviews.org, we constructed an undirected dynamic communication graph that contains 100 daily instances with the time span from November 8, 1997, to February 16, 1998. The largest graph among those instances has 3,569 nodes and 12,510 edges. The dataset shows both the addition and deletion of nodes and edges over time.

Power Grid. The second dataset is the power grid network. It is a static, undirected, unweighted network representing the topology of the Western States Power Grid of the United State [133], which has 4,941 nodes and 6,594 edges. To simulate the evolving process, we randomly add $0.5\%m$ (m is the number of edges in the graph) new edges to the graph at each time stamp as perturbation edges. We have changed different percentages of perturbation edges and experimented with several runs on each of the settings. As the results are similar, we only report the results from one run for brevity.

Airport. The third dataset is a static, undirected, unweighted airport network, which represents the internal U.S. air traffic lines between 2,649 airports and has 13,106 links (available

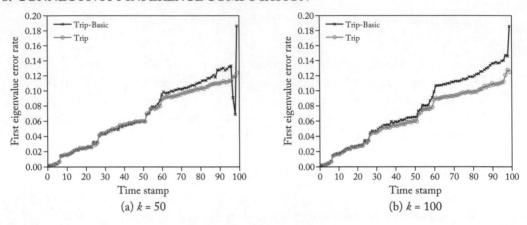

Figure 3.2: The error rate of first eigenvalue approximation.

at http://www.levmuchnik.net/Content/Networks/NetworkData.html). Again, a similar synthetic evolving process was done on this dataset. With similar experiment results, we only report those from one run of simulations for brevity.

Evaluation Metrics. For the quality of eigen-functions tracking, we use the error rate ϵ. For eigenvalues, the number of triangles and robustness measurement, their error rate are computed as $\epsilon = \frac{|g-g^*|}{g^*}$, where g and g^* are the estimated and true eigen-function values, respectively. For eigenvector, the error is computed as $\epsilon = 1 - \frac{\mathbf{u}\mathbf{u}^*}{\|\mathbf{u}\|\|\mathbf{u}^*\|}$, where $\mathbf{u}$ is the estimated eigenvector and $\mathbf{u}^*$ is the corresponding true eigenvector. For attribution analysis, we use the top-10 precision. For efficiency, we report the speedup of our algorithms over the re-computing strategy which computes the corresponding eigen-pairs from scratch at each time stamp.

Effectiveness Results

Effectiveness of Eigen-Function Tracking. Figures 3.2–3.6 compare the effectiveness of TRIP-BASIC and TRIP using different number of eigen-pairs (k). We have the following observations. First, for all of the four eigen-functions, both algorithms could reach an overall error rate below 20% at the end of the tracking process. Second, when k is increased from 50–100, TRIP-BASIC could get a relatively more stable approximation over the tracking process. Third, TRIP is more stable and overall reaches a smaller error rate compared with TRIP-BASIC. For example, as time goes by, TRIP-BASIC starts to fluctuate sharply when $k = 50$ on all four eigen-functions. Finally, the error on the number of triangles is relatively higher. This is probably because the number of triangles is the sum of cubic eigenvalues, and small errors on eigenvalues would therefore be magnified on the final result.

In addition, we also compared our algorithms with three different eigen-pair estimation methods, which include (1) "QR Decom," a QR decomposition-based eigen-pairs upda

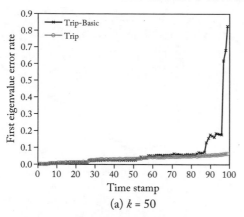
(a) $k = 50$

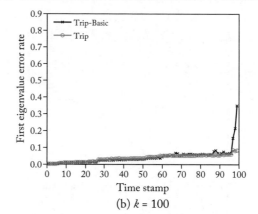
(b) $k = 100$

Figure 3.3: The error rate of first eigenvector approximation.

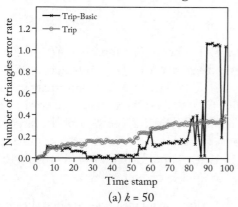
(a) $k = 50$

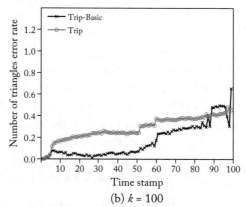
(b) $k = 100$

Figure 3.4: The error rate of number of triangles approximation.

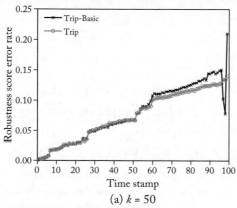
(a) $k = 50$

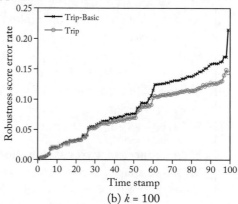
(b) $k = 100$

Figure 3.5: The error rate of robustness score approximation.

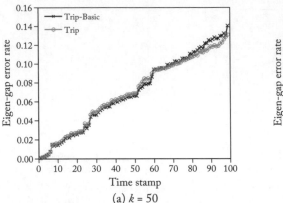 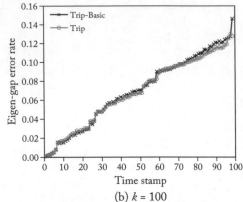

(a) $k = 50$ (b) $k = 100$

Figure 3.6: The error rate of eigen-gap approximation.

ing method [73]; (2) "SVD delta," simple SVD decomposition on $\Delta \mathbf{A}$; and (3) "Nystrom," a sampling-based eigen-pair estimation method derived from Nystrom algorithm [34]. For better effectiveness/efficiency trade-off, we sample 2,000 nodes for the Nystrom algorithm to calculate eigen-pairs in our experiment. To better illustrate the results, we take the error rates of all methods for every 15 days on the *AS* data set. As the "SVD delta" method causes large tracking errors compared to other methods, we only report the error rates from other comparing methods as shown from Figures 3.7–3.11. We can see that the performance of TRIP-BASIC and TRIP are among the best methods although their error rates keep increasing as time accumulates.

Effectiveness of Attribution Analysis. For attribution analysis, we divided the changed edges at each time stamp into two classes: edges being added and edges being removed. And among these two classes, we rank those edges according to their attribution score defined in Section 3.1.2. As a consequence, the top-ranked edges are the ones that have the largest impact on the corresponding eigen-functions. Here, we scored and ranked those edges with our approximated eigen-pairs and true eigen-pairs, respectively, and then compare the similarity between the two ranks. The precision of attribution analysis, therefore, is defined as the precision at rank 10 in the approximated rank list. As similar results are observed in all three data set, we only report those on *AS* dataset; as shown in Figures 3.12 and 3.13. For the analysis on both added edges and removed edges, TRIP overall outperforms TRIP-BASIC.

Effectiveness of Error Estimation. To show the effectiveness of Algorithm 3.4, we compare the curve shapes between true errors of TRIP and accumulative estimated errors $err_{acc}(\lambda^t)$ on *AS* data set with $k = 50$. Ideally, the two curves should overlap with each other when $err_{acc}(\lambda^t)$ is properly scaled with some elaborately picked factor. Figure 3.14 shows that the estimated error $err_{acc}(\lambda^t)$ can effectively catch sharp error increases in the tracking process as marked

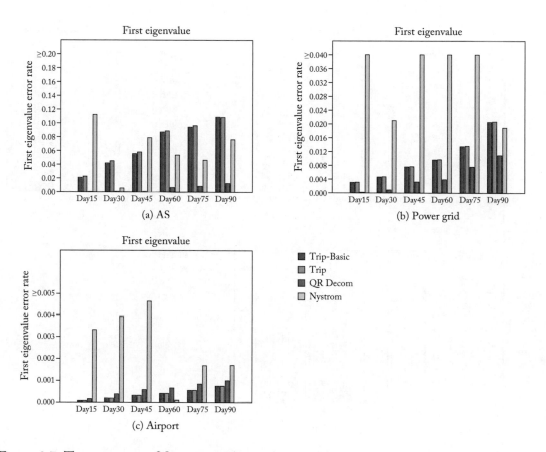

Figure 3.7: The error rate of first eigenvalue approximation.

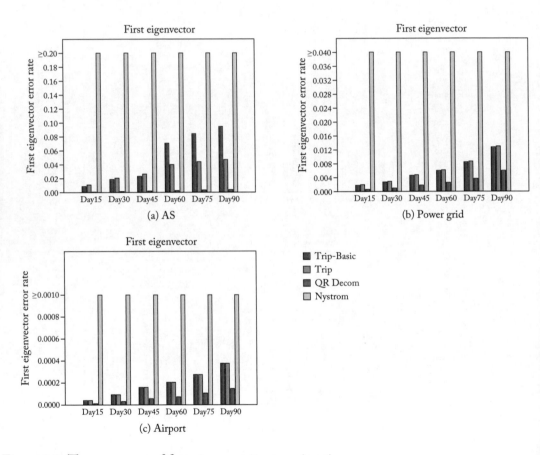

Figure 3.8: The error rate of first eigenvector approximation.

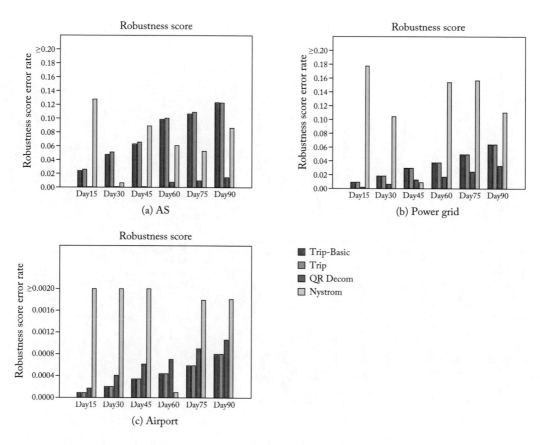

Figure 3.9: The error rate of robustness score approximation.

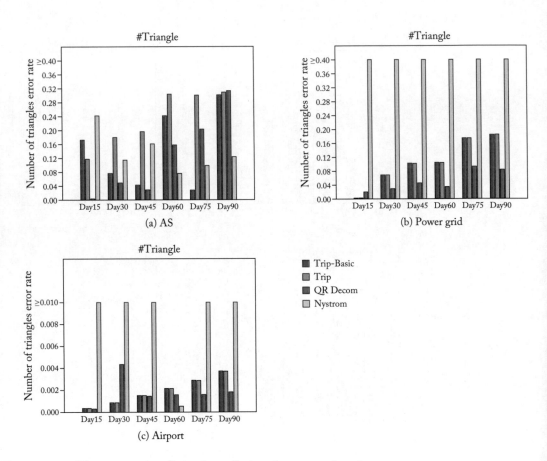

Figure 3.10: The error rate of number of triangles approximation.

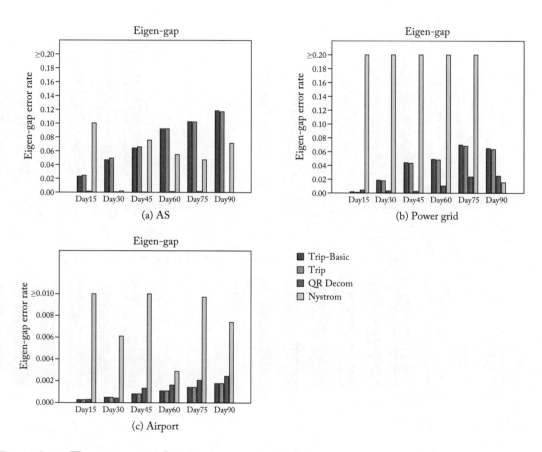

Figure 3.11: The error rate of eigen-gap approximation.

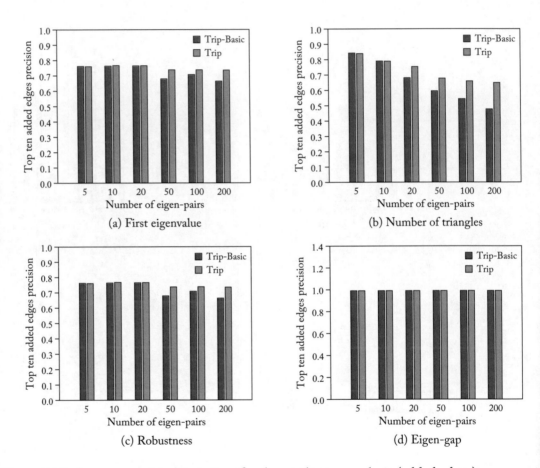

Figure 3.12: Average precision over time for the attribution analysis (added edges).

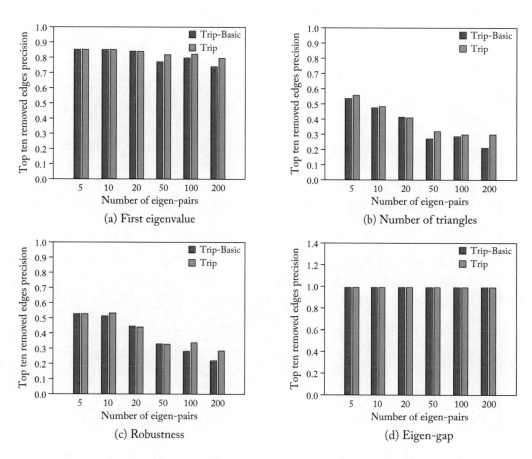

Figure 3.13: Average precision over time of the attribution analysis (removed edges).

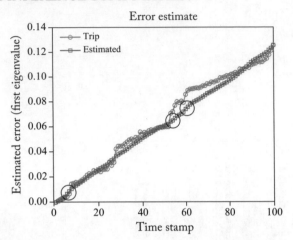

Figure 3.14: The estimated error of TRIP-BASIC and TRIP on **AS** data set.

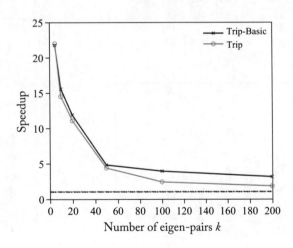

Figure 3.15: The running time speedup of TRIP-BASIC and TRIP w.e.t. to k.

in red circle. Therefore, it can be used as a trigger to restart the tracking process so that the accumulative error can always be kept within a low range.

Efficiency Results

Figure 3.15 shows the average speed up w.r.t. different k values on *AS* dataset. We see that both TRIP-BASIC and TRIP can achieve more than 20× speed up when k is small. As k increases, the speedup decreases.

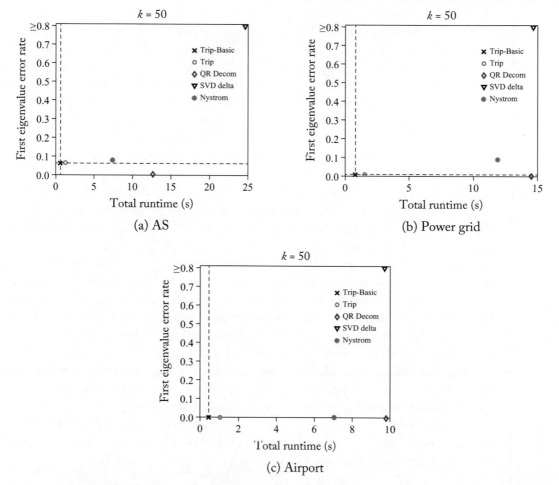

Figure 3.16: The error rate vs. total runtime of first eigenvalue approximation in 100 time stamps.

To further demonstrate the efficiency of the proposed algorithms, we also compare their effectiveness/efficiency trade-offs with those of the alternative methods. Figure 3.16 shows that our algorithms can keep the average error rate very small on all three data sets while consuming the least amount of time.

3.2   CROSS-LAYER DEPENDENCY INFERENCE

The interactions observed across different domains have facilitated the emergence of multi-layered networks. Examples of such complicated network systems include critical infrastructure networks mentioned in Chapter 1, collaboration platforms, and biological systems. One crucial topological structure that differentiates multi-layered networks from other network models is its *cross-layer dependency*, which describes the associations/dependencies between the nodes from different layers. For example, in infrastructure networks, the full functioning of the AS network depends on the sufficient power supply from the power grid layer, which in turn relies on the functioning of the transportation network (e.g., to deliver sufficient fuel). Similarly, in the biological systems, the dependency is represented as the associations among diseases, genes, and drugs. In practice, the cross-layer dependency often plays a central role in many multi-layered network mining tasks. For example, in the critical infrastructure network, the intertwined cross-layer dependency is considered as a major factor of system robustness. This is because a small failure on the supporting network (e.g., power station malfunction in the power grid) may be amplified in all its dependent networks through cross-layer dependencies, resulting in a catastrophic/cascading failure of the entire system. On the other hand, the cross-layer dependency in the biological system is often the key to new discoveries, such as new treatment associations between existing drugs and new diseases.

In spite of its key importance, it remains a daunting task to know the exact cross-layer dependency structure in a multi-layered network, due to noise, incomplete data sources, and limited accessibility to network dynamics. For example, an extreme weather event might significantly disrupt the power grid, the transportation network, and the cross-layer dependencies in between at the epicenter. Yet, due to limited accessibility to the damaged area during or soon after the disruption, the cross-layer dependency structure might only have a probabilistic and/or coarse-grained description. On the other hand, for a newly identified chemical in the biological system, its cross-layer dependencies w.r.t. proteins and/or the diseases might be completely unknown due to clinical limitations. (i.e., the *zero-start* problem).

In this work, we aim to tackle the above challenges by developing effective and efficient methods to infer cross-layer dependency on multi-layered networks. The main contributions of the work can be summarized as follows.

- *Problem Formulations:* We define the cross-layer dependency inference problem as a regularized optimization problem. The key idea behind this formulation is to collectively leverage the within-layer topology and the observed cross-layer dependency to infer a latent, low-rank representation for each layer, which can be used to infer the missing cross-layer dependencies in the network.

- *Algorithms and Analysis:* We propose an effective algorithm—FASCINATE to infer the cross-layer dependency on multi-layered networks, and analyze its optimality, convergence, and complexity. We further present its variants and generalizations, including an online algorithm to address the *zero-start* problem.

- *Evaluations:* We perform extensive experiments on five real datasets to substantiate the effectiveness, efficiency, and scalability of our proposed algorithms. Specifically, our experimental evaluations show that the proposed algorithms outperform their best competitors

Table 3.2: Main symbols FASCINATE.

Symbol	Definition and Description
A, B	the adjacency matrices (bold upper case)
a, b	column vectors (bold lower case)
$\mathcal{A}, \mathcal{B}$	sets (calligraphic)
$\mathbf{A}(i,j)$	the element at ith row jth column in matrix $\mathbf{A}$
$\mathbf{A}(i,:)$	the ith row of matrix $\mathbf{A}$
$\mathbf{A}(:,j)$	the jth column of matrix $\mathbf{A}$
$\mathbf{A}'$	transpose of matrix $\mathbf{A}$
$\hat{\mathbf{A}}$	the adjacency matrix of $\mathbf{A}$ with the newly added node
$\mathbf{G}$	the layer-layer dependency matrix
$\mathcal{A}$	within-layer connectivity matrices of the network $\mathcal{A} = \{\mathbf{A}_1, \ldots, \mathbf{A}_g\}$
$\mathcal{D}$	cross-layer dependency matrices $\mathcal{D} = \{\mathbf{D}_{i,j}\ i,j = 1, \ldots, g\}$
$\mathbf{W}_{i,j}$	weight matrix for $\mathbf{D}_{i,j}$
$\mathbf{F}_i$	low-rank representation for layer-i ($i = 1, \ldots, g$)
m_i, n_i	number of edges and nodes in graph $\mathbf{A}_i$
$m_{i,j}$	number of dependencies in $\mathbf{D}_{i,j}$
g	total number of layers
r	the rank for $\{\mathbf{F}_i\}_{i=1,\ldots,g}$
t	the maximal iteration number
ξ	the threshold to determine the iteration

by 8.2–41.9% in terms of inference accuracy while enjoying linear complexity. Moreover, the proposed FASCINATE-ZERO algorithm can achieve up to $10^7 \times$ speedup with barely any compromise on accuracy.

3.2.1   PROBLEM DEFINITION

In this section, we give the formal definitions of the cross-layer dependency inference problems. The main symbols used in this work are listed in Table 3.2. Following the convention, we use bold uppercase for matrices (e.g., $\mathbf{A}$), bold lowercase for vectors (e.g., $\mathbf{a}$), and calligraphic for sets (e.g., $\mathcal{A}$). $\mathbf{A}'$ denotes the transpose of matrix $\mathbf{A}$. We use the ˆ sign to denote the notations after a new node is accommodated to the system (e.g., $\hat{J}$, $\hat{\mathbf{A}}_1$), and the ones without the $\hat{\ }$ sign as the notations before the new node arrives.

While several multi-layered network models exist in the literature, we will focus on a recent model proposed in [15], due to its flexibility to model more complicated, cross-layer dependency structure. We refer the readers to Chapter 2 for its full details. For the purpose of this work, we mainly need the following notations to describe a multi-layered network with g layers. First, we need a $g \times g$ layer-layer dependency matrix $\mathbf{G}$, where $\mathbf{G}(i, j) = 1$ if layer-j depends on layer-i, and $\mathbf{G}(i, j) = 0$ otherwise. Second, we need a set of g within-layer connectivity matrices: $\mathcal{A} = \{\mathbf{A}_1, ..., \mathbf{A}_g\}$ to describe the connectivities/similarities between nodes within the same layer. Third, we need a set of cross-layer dependency matrices $\mathcal{D} = \{\mathbf{D}_{i,j} \ i, j = 1, ..., g\}$, where $\mathbf{D}_{i,j}$ describes the dependencies between the nodes from layer-i and the nodes from layer-j if these two layers are directly dependent (i.e., $\mathbf{G}(i, j) = 1$). When there are no direct dependencies between the two layers (i.e., $\mathbf{G}(i, j) = 0$), the corresponding dependency matrix $\mathbf{D}_{i,j}$ is absent. Taking the multi-layered network in Figure 3.17 as an example, the abstract layer-layer dependency network $\mathbf{G}$ of this biological system can be viewed as a line graph. The four within-layer similarity matrices in $\mathcal{A}$ are the *chemical network* $(\mathbf{A}_1)$, the *drug network* $(\mathbf{A}_2)$, the *disease network* $(\mathbf{A}_3)$, and the *protein-protein interaction (PPI) network* $(\mathbf{A}_4)$. Across those layers, we have three non-empty dependency matrices, including the *chemical-drug* dependency matrix $(\mathbf{D}_{1,2})$, the *drug-disease* interaction matrix $(\mathbf{D}_{2,3})$ and the *disease-protein* dependency matrix $(\mathbf{D}_{3,4})$.[2]

As mentioned earlier, it is often very hard to accurately know the cross-layer dependency matrices $\{\mathbf{D}_{i,j} \ i, j = 1, ..., g\}$. In other words, such *observed* dependency matrices are often incomplete and noisy. Inferring the missing cross-layer dependencies is an essential prerequisite for many multi-layered network mining tasks. On the other hand, real-world networks are evolving over time. Probing the cross-layer dependencies is often a time-consuming process in large complex networks. Thus, a newly added node could have no observed cross-layer dependencies for a fairly long period of time since its arrival. Therefore, inferring the dependencies of such kind of *zero-start* nodes is an important problem that needs to be solved efficiently. Formally, we define the cross-layer dependency inference problem (CODE) and its corresponding *zero-start* variant (CODE-ZERO) as follows.

Problem 3.4   (Code) Cross-Layer Dependency Inference

Given: A multi-layered network with (1) layer-layer dependency matrix $\mathbf{G}$; (2) within-layer connectivity matrices $\mathcal{A} = \{\mathbf{A}_1, ..., \mathbf{A}_g\}$; and (3) observed cross-layer dependency matrices $\mathcal{D} = \{\mathbf{D}_{i,j} \ i, j = 1, ..., g\}$.

Output: The true cross-layer dependency matrices $\{\tilde{\mathbf{D}}_{i,j} \ i, j = 1, ..., g\}$.

Problem 3.5   (Code-ZERO) Cross-Layer Dependency Inference for *zero-start* Nodes

---

[2]More complicated dependency relationships may exist across the layers in real settings, which can be addressed with our model as well.

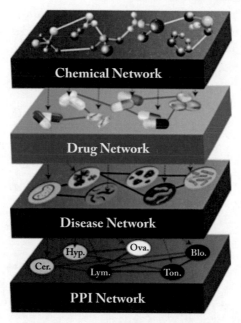

Figure 3.17: A simplified four-layered network for biological systems.

Given: (1) A multi-layered network $\{\mathbf{G}, \mathcal{A}, \mathcal{D}\}$; (2) a newly added node p in the lth layer; and (3) a $1 \times n_l$ vector $\mathbf{s}$ that records the connections between p and the existing n_l nodes in layer l.

Output: The true dependencies between node p and nodes in dependent layers of layer-l, i.e., $\tilde{\mathbf{D}}_{l,j}(p,:)$ ($j = 1, ..., g$, $\mathbf{G}(l, j) = 1$).

3.2.2 PROPOSED ALGORITHMS FOR CODE

In this section, we present our proposed solution for Problem 3.4 (CODE). We start with the proposed optimization formulation and then present our algorithm (FASCINATE), followed by some effectiveness and efficiency analysis.

Fascinate: Optimization Formulation

The key idea behind our formulation is to treat Problem 3.4 as a *collective collaborative filtering* problem. To be specific, if we view (1) nodes from a given layer (e.g., power plants) as objects from a given domain (e.g., users/items), (2) the within-layer connectivity (e.g., communication networks) as an object-object similarity measure, and (3) the cross-layer dependency (e.g., dependencies between computers in the communication layer and the power plants in the power grid layer) as the "ratings" from objects of one domain to those of another domain; then inferring the missing cross-layer dependencies can be viewed as a task of inferring the missing ratings between the objects (e.g., users, items) across different domains. Having this analogy in mind, we propose to formulate Problem 3.4 as the following regularized optimization problem:

$$\min_{\mathbf{F}_i \geq 0 (i=1,\dots,g)} J = \underbrace{\sum_{i,j:\ \mathbf{G}(i,j)=1} \|\mathbf{W}_{i,j} \odot (\mathbf{D}_{i,j} - \mathbf{F}_i \mathbf{F}_j')\|_F^2}_{\text{C1: Matching Observed Cross-Layer Dependencies}} \qquad (3.9)$$

$$+ \underbrace{\alpha \sum_{i=1}^{g} \text{tr}(\mathbf{F}_i'(\mathbf{T}_i - \mathbf{A}_i)\mathbf{F}_i)}_{\text{C2: Node Homophily}} + \underbrace{\beta \sum_{i=1}^{g} \|\mathbf{F}_i\|_F^2}_{\text{C3: Regularization}},$$

where $\mathbf{T}_i$ is the diagonal degree matrix of $\mathbf{A}_i$ with $\mathbf{T}_i(u,u) = \sum_{v=1}^{n_i} \mathbf{A}_i(u,v)$; $\mathbf{W}_{i,j}$ is an $n_i \times n_j$ weight matrix to assign different weights to different entries in the corresponding cross-layer dependency matrix $\mathbf{D}_{i,j}$; and $\mathbf{F}_i$ is the low-rank representation for layer i. For now, we set the weight matrices as follows: $\mathbf{W}_{i,j}(u,v)$
$= 1$ if $\mathbf{D}_{i,j}(u,v)$ is observed, and $\mathbf{W}_{i,j}(u,v) \in [0,1]$ if $\mathbf{D}_{i,j}(u,v)$
$= 0$ (i.e., unobserved). To simplify the computation, we set the weights of all unobserved entries to a global value w. We will discuss alternative choices for the weight matrices as the variants of FASCINATE later this section.

In this formulation (Eq. (3.9)), we can think of $\mathbf{F}_i$ as the low-rank representations/features of the nodes in layer i in some latent space, which is shared among different layers. The cross-layer dependencies between the nodes from two dependent layers can be viewed as the inner product of their latent features. Therefore, the intuition of the first term (i.e., C1) is that we want to match all the cross-layer dependencies, calibrated by the weight matrix $\mathbf{W}_{i,j}$. The second term (i.e., C2) is used to achieve node homophily, which says that for a pair of nodes u and v from the same layer (say layer-i), their low-rank representations should be similar (i.e., small $\|\mathbf{F}_i(u,:) - \mathbf{F}_i(v,:)\|_2$) if the within-layer connectivity between these two nodes is strong (i.e., large $\mathbf{A}_i(u,v)$). The third term (i.e., C3) is to regularize the norm of the low-rank matrices $\{\mathbf{F}_i\}_{i=1,\dots,g}$ to prevent over-fitting.

Once we solve Eq. (3.9), for a given node u from layer-i and a node v from layer-j, the cross-layer dependency between them can be estimated as $\tilde{\mathbf{D}}_{i,j}(u,v) = \mathbf{F}_i(u,:)\mathbf{F}_j(v,:)'$.

Fascinate: Optimization Algorithm

The optimization problem defined in Eq. (3.9) is non-convex. Thus, we seek to find a local optima by the block coordinate descent method, where each $\mathbf{F}_i$ naturally forms a "block." To be specific, if we fix all other $\mathbf{F}_j (j = 1, \dots, g, j \neq i)$ and ignore the constant terms, Eq. (3.9) ca be simplified as

$$J_i = \sum_{j:\, G(i,j)=1} \|\mathbf{W}_{i,j} \odot (\mathbf{D}_{i,j} - \mathbf{F}_i \mathbf{F}'_j)\|_F^2 + \alpha \mathrm{tr}(\mathbf{F}'_i (\mathbf{T}_i - \mathbf{A}_i)\mathbf{F}_i) + \beta \|\mathbf{F}_i\|_F^2. \qquad (3.10)$$

The derivative of J_i w.r.t. $\mathbf{F}_i$ is

$$\frac{\partial J_i}{\partial \mathbf{F}_i} = 2\Bigg(\sum_{j:\, G(i,j)=1} [-(\mathbf{W}_{i,j} \odot \mathbf{W}_{i,j} \odot \mathbf{D}_{i,j})\mathbf{F}_j + (\mathbf{W}_{i,j} \odot \mathbf{W}_{i,j} \odot (\mathbf{F}_i \mathbf{F}'_j))\mathbf{F}_j] \qquad (3.11)$$

$$+ \alpha \mathbf{T}_i \mathbf{F}_i - \alpha \mathbf{A}_i \mathbf{F}_i + \beta \mathbf{F}_i \Bigg).$$

A fixed-point solution of Eq. (3.11) with non-negativity constraint on $\mathbf{F}_i$ leads to the following multiplicative updating rule for $\mathbf{F}_i$

$$\mathbf{F}_i(u, v) \leftarrow \mathbf{F}_i(u, v) \sqrt{\frac{\mathbf{X}(u, v)}{\mathbf{Y}(u, v)}}, \qquad (3.12)$$

where

$$\mathbf{X} = \sum_{j:\, G(i,j)=1} (\mathbf{W}_{i,j} \odot \mathbf{W}_{i,j} \odot \mathbf{D}_{i,j})\mathbf{F}_j + \alpha \mathbf{A}_i \mathbf{F}_i \qquad (3.13)$$

$$\mathbf{Y} = \sum_{j:\, G(i,j)=1} (\mathbf{W}_{i,j} \odot \mathbf{W}_{i,j} \odot (\mathbf{F}_i \mathbf{F}'_j))\mathbf{F}_j + \alpha \mathbf{T}_i \mathbf{F}_i + \beta \mathbf{F}_i.$$

Recall that we set $\mathbf{W}_{i,j}(u, v) = 1$ when $\mathbf{D}_{i,j}(u, v) > 0$, and $\mathbf{W}_{i,j}(u, v) = w$ when $\mathbf{D}_{i,j}(u, v) = 0$. Here, we define $\mathbf{I}^O_{i,j}$ as an indicator matrix for the observed entries in $\mathbf{D}_{i,j}$, that is, $\mathbf{I}^O_{i,j}(u, v) = 1$ if $\mathbf{D}_{i,j}(u, v) > 0$, and $\mathbf{I}^O_{i,j}(u, v) = 0$ if $\mathbf{D}_{i,j}(u, v) = 0$. Then, the estimated dependencies over the observed data can be represented as $\tilde{\mathbf{R}}_{i,j} = \mathbf{I}^O_{i,j} \odot (\mathbf{F}_i \mathbf{F}_j)$. With these notations, we can further simplify the update rule in Eq. (3.13) as follows:

$$\mathbf{X} = \sum_{j:\, G(i,j)=1} \mathbf{D}_{i,j} \mathbf{F}_j + \alpha \mathbf{A}_i \mathbf{F}_i \qquad (3.14)$$

$$\mathbf{Y} = \sum_{j:\, G(i,j)=1} ((1 - w^2)\tilde{\mathbf{R}}_{i,j} + w^2 \mathbf{F}_i \mathbf{F}'_j)\mathbf{F}_j + \alpha \mathbf{T}_i \mathbf{F}_i + \beta \mathbf{F}_i. \qquad (3.15)$$

The proposed FASCINATE algorithm is summarized in Algorithm 3.5. First, it randomly initializes the low-rank matrices for each layer (steps 1–3). Then, it begins the iterative update procedure. In each iteration (steps 4–10), the algorithm alternatively updates $\{\mathbf{F}_i\}_{i=1,\dots,g}$ one by one. We use two criteria to terminate the iteration: (1) either the Frobenius norm between two successive iterations for all $\{\mathbf{F}_i\}_{i=1,\dots,g}$ is less than a threshold ξ, or (2) the maximum iteration number t is reached.

Algorithm 3.5 The FASCINATE Algorithm

Input: (1) A multi-layered network with (a) layer-layer dependency matrix $\mathbf{G}$, (b) within-layer connectivity matrices $\mathcal{A} = \{\mathbf{A}_1, ..., \mathbf{A}_g\}$, and (c) observed cross-layer node dependency matrices $\mathcal{D} = \{\mathbf{D}_{i,j} \ i, j = 1, ..., g\}$; (2) the rank size r; (3) weight w; and (4) regularized parameters α and β

Output: Low-rank representations for each layer $\{\mathbf{F}_i\}_{i=1,...,g}$

1: **for** $i = 1$ to g **do**
2:    initialized $\mathbf{F}_i$ as $n_i \times r$ non-negative random matrix
3: **end for**
4: **while** not converge **do**
5:    **for** $i = 1$ to g **do**
6:       compute $\mathbf{X}$ as Eq. (3.14)
7:       compute $\mathbf{Y}$ as Eq. (3.15)
8:       update $\mathbf{F}_i$ as Eq. (3.12)
9:    **end for**
10: **end while**
11: return $\{\mathbf{F}_i\}_{i=1,...,g}$

Proof and Analysis

Here, we analyze the proposed FASCINATE algorithm in terms of its effectiveness as well as its efficiency.

Effectiveness Analysis. In terms of effectiveness, we show that the proposed FASCINATE algorithm indeed finds a local optimal solution to Eq. (3.9). To see this, we first give the following theorem, which says that the fixed point solution of Eq. (3.12) satisfies the *KKT* condition.

Theorem 3.6   *The fixed point solution of Eq. (3.12) satisfies the KKT condition.*

Proof. The Lagrangian function of Eq. (3.10) can be written as

$$L_i = \sum_{j: \ \mathbf{G}(i,j)=1} \|\mathbf{W}_{i,j} \odot (\mathbf{D}_{i,j} - \mathbf{F}_i \mathbf{F}_j')\|_F^2 \tag{3.16}$$

$$+ \alpha \mathrm{tr}(\mathbf{F}_i'\mathbf{T}_i\mathbf{F}_i) - \alpha \mathrm{tr}(\mathbf{F}_i'\mathbf{A}_i\mathbf{F}_i) + \beta \|\mathbf{F}_i\|_F^2 - \mathrm{tr}(\mathbf{\Lambda}'\mathbf{F}_i),$$

where $\mathbf{\Lambda}$ is the Lagrange multiplier. Setting the derivative of L_i w.r.t. $\mathbf{F}_i$ to 0, we get

$$2 \ (\sum_{j: \ \mathbf{G}(i,j)=1} [-(\mathbf{W}_{i,j} \odot \mathbf{W}_{i,j} \odot \mathbf{D}_{i,j})\mathbf{F}_j + (\mathbf{W}_{i,j} \odot \mathbf{W}_{i,j} \odot (\mathbf{F}_i\mathbf{F}_j'))\mathbf{F}_j] \tag{3.17}$$

$$+ \alpha \mathbf{T}_i\mathbf{F}_i - \alpha \mathbf{A}_i\mathbf{F}_i + \beta \mathbf{F}_i) = \mathbf{\Lambda}.$$

By the KKT complementary slackness condition, we have

$$\Big[\underbrace{\sum_{j:\ G(i,j)=1} (\mathbf{W}_{i,j} \odot \mathbf{W}_{i,j} \odot (\mathbf{F}_i \mathbf{F}_j{}'))\mathbf{F}_j + \alpha \mathbf{T}_i \mathbf{F}_i + \beta \mathbf{F}_i}_{\mathbf{Y}} \tag{3.18}$$

$$- \underbrace{(\sum_{j:\ G(i,j)=1} (\mathbf{W}_{i,j} \odot \mathbf{W}_{i,j} \odot \mathbf{D}_{i,j})\mathbf{F}_j + \alpha \mathbf{A}_i \mathbf{F}_i)}_{\mathbf{X}} \Big](u,v)\mathbf{F}_i(u,v) = 0.$$

Therefore, we can see that the fixed point solution of Eq. (3.12) satisfies the above equation.
□

The convergence of the proposed FASCINATE algorithm is given by the following lemma.

Lemma 3.7   *Under the updating rule in Eq. (3.12), the objective function in Eq. (3.10) decreases monotonically.*

Proof. By expending the Frobenius norms and dropping constant terms, Eq. (3.10) can be further simplified as

$$J_i = \sum_{G_{(i,j)}=1} \underbrace{(-2\mathrm{tr}((\mathbf{W}_{i,j} \odot \mathbf{W}_{i,j} \odot \mathbf{D}_{i,j})\mathbf{F}_j \mathbf{F}_i')}_{T_1} + \underbrace{\mathrm{tr}((\mathbf{W}_{i,j} \odot \mathbf{W}_{i,j} \odot (\mathbf{F}_i \mathbf{F}_j'))\mathbf{F}_j \mathbf{F}_i'))}_{T_2} \tag{3.19}$$

$$+ \underbrace{\alpha \mathrm{tr}(\mathbf{F}_i' \mathbf{T}_i \mathbf{F}_i)}_{T_3} \underbrace{-\alpha \mathrm{tr}(\mathbf{F}_i' \mathbf{A} \mathbf{F}_i)}_{T_4} + \underbrace{\beta \mathrm{tr}(\mathbf{F}_i \mathbf{F}_i')}_{T_5}.$$

Following the auxiliary function approach in [63], the auxiliary function $H(\mathbf{F}_i, \tilde{\mathbf{F}}_i)$ of J_i must satisfy

$$H(\mathbf{F}_i, \mathbf{F}_i) = J_i, \quad H\left(\mathbf{F}_i, \tilde{\mathbf{F}}_i\right) \geq J_i. \tag{3.20}$$

Define

$$\mathbf{F}_i^{(t+1)} = \arg\min_{\mathbf{F}_i} H\left(\mathbf{F}_i, \mathbf{F}_i^{(t)}\right). \tag{3.21}$$

By this construction, we have

$$J_i^{(t)} = H\left(\mathbf{F}_i^{(t)}, \mathbf{F}_i^{(t)}\right) \geq H\left(\mathbf{F}_i^{(t+1)}, \mathbf{F}_i^{(t)}\right) \geq J_i^{(t+1)}, \tag{3.22}$$

which proves that $J_i^{(t)}$ decreases monotonically.

Next, we prove that (1) we can find an auxiliary function that satisfies the above constraints and (2) the updating rule in Eq. (3.12) leads to the global minimum solution to the auxiliary function.

First, we show that the following function is one of the auxiliary functions of Eq. (3.19):

$$H\left(\mathbf{F}_i, \tilde{\mathbf{F}}_i\right) = \sum_{\mathbf{G}_{(i,j)}=1} (T_1' + T_2') + T_3' + T_4' + T_5', \tag{3.23}$$

where

$$T_1' = -2 \sum_{u=1}^{n_i} \sum_{k=1}^{r} [(\mathbf{W}_{i,j} \odot \mathbf{W}_{i,j} \odot \mathbf{D}_{i,j})\mathbf{F}_j](u,k)\tilde{\mathbf{F}}_i(u,k) \left(1 + \log \frac{\mathbf{F}_i(u,k)}{\tilde{\mathbf{F}}_i(u,k)}\right) \tag{3.24}$$

$$T_2' = \sum_{u=1}^{n_i} \sum_{k=1}^{r} \frac{[(\mathbf{W}_{i,j} \odot \mathbf{W}_{i,j} \odot (\tilde{\mathbf{F}}_i \mathbf{F}_j'))\mathbf{F}_j](u,k)\mathbf{F}_i^2(u,k)}{\tilde{\mathbf{F}}_i(u,k)} \tag{3.25}$$

$$T_3' = \sum_{u=1}^{n_i} \sum_{k=1}^{r} \frac{[\alpha\mathbf{T}_i\tilde{\mathbf{F}}_i](u,k)\mathbf{F}_i^2(u,k)}{\tilde{\mathbf{F}}_i(u,k)} \tag{3.26}$$

$$T_4' = -\sum_{u=1}^{n_i} \sum_{v=1}^{n_i} \sum_{k=1}^{r} \alpha\mathbf{A}_i(u,v)\tilde{\mathbf{F}}_i(v,k)\tilde{\mathbf{F}}_i(u,k) \left(1 + \log \frac{\mathbf{F}_i(v,k)\mathbf{F}_i(u,k)}{\tilde{\mathbf{F}}_i(v,k)\tilde{\mathbf{F}}_i(u,k)}\right) \tag{3.27}$$

$$T_5' = \sum_{u=1}^{n_i} \sum_{k=1}^{r} \beta\mathbf{F}_i^2(u,k). \tag{3.28}$$

Here, we prove that $T_i' \geq T_i$ for $i = 1, \ldots, 5$ term by term.

Using the inequality $z \geq 1 + \log z$, we have

$$T_1' \geq -2 \sum_{u=1}^{n_i} \sum_{k=1}^{r} [(\mathbf{W}_{i,j} \odot \mathbf{W}_{i,j} \odot \mathbf{D}_{i,j})\mathbf{F}_j](u,k)\mathbf{F}_i(u,k) = T_1 \tag{3.29}$$

$$T_4' \geq -\sum_{u=1}^{n_i} \sum_{v=1}^{n_i} \sum_{k=1}^{r} \alpha\mathbf{A}_i(u,v)\mathbf{F}_i(v,k)\mathbf{F}_i(u,k) = T_4.$$

Expanding T_2', we can rewrite it as

$$T_2' = \sum_{u=1}^{n_i} \sum_{v=1}^{n_j} \sum_{k=1}^{r} \sum_{l=1}^{r} \frac{\mathbf{W}_{i,j}^2(u,v)\tilde{\mathbf{F}}_i(u,l)\mathbf{F}_j'(l,v)\mathbf{F}_j(v,k)\mathbf{F}_i^2(u,k)}{\tilde{\mathbf{F}}_i(u,k)}. \tag{3.30}$$

Let $\mathbf{F}_i(u,k) = \tilde{\mathbf{F}}_i(u,k)\mathbf{Q}_i(u,k)$, then

$$T_2' = \sum_{u=1}^{n_i} \sum_{v=1}^{n_j} \sum_{k=1}^{r} \sum_{l=1}^{r} \mathbf{W}_{i,j}^2(u,v)\mathbf{F}_j'(l,v)\mathbf{F}_j(v,k)\tilde{\mathbf{F}}_i(u,l)\tilde{\mathbf{F}}_i(u,k)\mathbf{Q}_i^2(u,k) \qquad (3.31)$$

$$= \sum_{u=1}^{n_i} \sum_{v=1}^{n_j} \sum_{k=1}^{r} \sum_{l=1}^{r} \mathbf{W}_{i,j}^2(u,v)\mathbf{F}_j'(l,v)\mathbf{F}_j(v,k)\tilde{\mathbf{F}}_i(u,l)\tilde{\mathbf{F}}_i(u,k)\left(\frac{\mathbf{Q}_i^2(u,k) + \mathbf{Q}_i^2(u,l)}{2}\right)$$

$$\geq \sum_{u=1}^{n_i} \sum_{v=1}^{n_j} \sum_{k=1}^{r} \sum_{l=1}^{r} \mathbf{W}_{i,j}^2(u,v)\mathbf{F}_j'(l,v)\mathbf{F}_j(v,k)\tilde{\mathbf{F}}_i(u,l)\tilde{\mathbf{F}}_i(u,k)\mathbf{Q}_i(u,k)\mathbf{Q}_i(u,l)$$

$$= \sum_{u=1}^{n_i} \sum_{v=1}^{n_j} \sum_{k=1}^{r} \sum_{l=1}^{r} \mathbf{W}_{i,j}^2(u,v)\mathbf{F}_i(u,l)\mathbf{F}_j'(l,v)\mathbf{F}_j(v,k)\mathbf{F}_i(u,k)$$

$$= T_2.$$

For T_3', by using the following inequality in [33]

$$\sum_{i=1}^{n} \sum_{p=1}^{k} \frac{[\mathbf{A}\mathbf{S}^*\mathbf{B}](i,p)\mathbf{S}^2(i,p)}{\mathbf{S}^*(i,p)} \geq \mathrm{tr}(\mathbf{S}'\mathbf{A}\mathbf{S}\mathbf{B}), \qquad (3.32)$$

where $\mathbf{A} \in \mathbb{R}_+^{n \times n}$, $\mathbf{B} \in \mathbb{R}_+^{k \times k}$, $\mathbf{S} \in \mathbb{R}_+^{n \times k}$, $\mathbf{S}^* \in \mathbb{R}_+^{n \times k}$, and $\mathbf{A}, \mathbf{B}$ are symmetric, we have

$$T_3' \geq \alpha \mathrm{tr}(\mathbf{F}_i'\mathbf{T}_i\mathbf{F}_i) = T_3. \qquad (3.33)$$

For T_5', we have $T_5' = T_5$. Putting the above inequalities together, we have $H(\mathbf{F}_i, \tilde{\mathbf{F}}_i) \geq J_i^s(\mathbf{F}_i)$. Next, we find the global minimum solution to $H(\mathbf{F}_i, \tilde{\mathbf{F}}_i)$. The gradient of $H(\mathbf{F}_i, \tilde{\mathbf{F}}_i)$ is

$$\frac{1}{2}\frac{\partial H(\mathbf{F}_i, \tilde{\mathbf{F}}_i)}{\partial \mathbf{F}_i(u,k)} = -\frac{[(\mathbf{W}_{i,j} \odot \mathbf{W}_{i,j} \odot \mathbf{D}_{i,j})\mathbf{F}_j](u,k)\tilde{\mathbf{F}}_i(u,k)}{\mathbf{F}_i(u,k)} \qquad (3.34)$$

$$+ \frac{[(\mathbf{W}_{i,j} \odot \mathbf{W}_{i,j} \odot (\tilde{\mathbf{F}}_i\mathbf{F}_j'))\mathbf{F}_j](u,k)\mathbf{F}_i(u,k)}{\tilde{\mathbf{F}}_i(u,k)}$$

$$+ \frac{[\alpha\mathbf{T}_i\tilde{\mathbf{F}}_i](u,k)\mathbf{F}_i(u,k)}{\tilde{\mathbf{F}}_i(u,k)} - \frac{[\alpha\mathbf{A}_i\tilde{\mathbf{F}}_i](u,k)\tilde{\mathbf{F}}_i(u,k)}{\mathbf{F}_i(u,k)} + \beta\mathbf{F}_i(u,k).$$

From the gradient of $H(\mathbf{F}_i, \tilde{\mathbf{F}}_i)$, we can easily get its Hessian matrix, which is a positive diagonal matrix. Therefore, the global minimum of $H(\mathbf{F}_i, \tilde{\mathbf{F}}_i)$ can be obtained by setting its gradient Eq. (3.34) to zero, which leads to

$$\mathbf{F}_i^2(u,k) = \tilde{\mathbf{F}}_i^2(u,k)\frac{[(\mathbf{W}_{i,j} \odot \mathbf{W}_{i,j} \odot (\tilde{\mathbf{F}}_i\mathbf{F}_j'))\mathbf{F}_j + \alpha\mathbf{A}_i\tilde{\mathbf{F}}_i](u,k)}{[(\mathbf{W}_{i,j} \odot \mathbf{W}_{i,j} \odot \mathbf{D}_{i,j})\mathbf{F}_j + \alpha\mathbf{T}_i\tilde{\mathbf{F}}_i + \beta\tilde{\mathbf{F}}_i](u,k)}. \qquad (3.35)$$

Recall that we have set $\mathbf{F}_i^{(t+1)} = \mathbf{F}_i$ and $\mathbf{F}_i^{(t)} = \tilde{\mathbf{F}}_i$. The above equation proves that the updating rule in Eq. (3.10) decreases monotonically. $\square$

According to Theorem 3.6 and Lemma 3.7, we conclude that Algorithm 3.5 converges to a local minima solution for Eq. (3.10) w.r.t. each individual $\mathbf{F}_i$.

Efficiency Analysis. In terms of efficiency, we analyze both the time complexity as well as the space complexity of the proposed FASCINATE algorithm, which are summarized in Lemmas 3.8 and 3.9. We can see that FASCINATE scales linearly w.r.t. the size of the entire multi-layered network.

Lemma 3.8   *The time complexity of Algorithm 3.5 is $O([\sum_{i=1}^{g} (\sum_{j:\, \mathbf{G}_{(i,j)}=1} (m_{i,j}r + (n_i + n_j)r^2) + m_i r)]t)$.*

Proof. In each iteration in Algorithm 3.5 for updating $\mathbf{F}_i$, the complexity of calculating $\mathbf{X}$ by Eq. (3.14) is $O(\sum_{j:\, \mathbf{G}(i,j)=1} m_{i,j}r + m_i r)$ due to the sparsity of $\mathbf{D}_{i,j}$ and $\mathbf{A}_i$. The complexity of computing $\tilde{\mathbf{R}}_{i,j}$ in $\mathbf{Y}$ is $O(m_{i,j}r)$. Computing $\mathbf{F}_i(\mathbf{F}_j'\mathbf{F}_j)$ requires $O((n_i + n_j)r^2)$ operations and computing $\alpha \mathbf{T}_i \mathbf{F}_i + \beta \mathbf{F}_i$ requires $O(n_i r)$ operations. So, it is of $O(\sum_{j:\, \mathbf{G}(i,j)=1} (m_{i,j}r + (n_i + n_j)r^2))$ complexity to get $\mathbf{Y}$ in step 7. Therefore, it takes $O(\sum_{j:\, \mathbf{G}(i,j)=1} (m_{i,j}r + (n_i + n_j)r^2) + m_i r)$ to update $\mathbf{F}_i$. Putting all together, the complexity of updating all low-rank matrices in each iteration is $O(\sum_{i=1}^{g} (\sum_{j:\, \mathbf{G}(i,j)=1} (m_{i,j}r + (n_i + n_j)r^2) + m_i r))$. Thus, the overall complexity of Algorithm 3.5 is $O([\sum_{i=1}^{g} (\sum_{\mathbf{G}(i,j)=1} (m_{i,j}r + (n_i + n_j)r^2) + m_i r)]t)$, where t is the maximum number of iterations in the algorithm. $\square$

Lemma 3.9   *The space complexity of Algorithm 3.5 is $O(\sum_{i=1}^{g}(n_i r + m_i) + \sum_{i,j:\, \mathbf{G}(i,j)=1} m_{i,j})$.*

Proof. It takes $O(\sum_{i=1}^{g} n_i r)$ to store all the low-rank matrices, and $O(\sum_{i=1}^{g} m_i + \sum_{i,j:\, \mathbf{G}(i,j)=1} m_{i,j})$ to store all the within-layer connectivity matrices and dependency matrices in the multi-layered network. To calculate $\mathbf{X}$ for $\mathbf{F}_i$, it costs $O(n_i r)$ to compute $\sum_{j:\, \mathbf{G}(i,j)=1} \mathbf{D}_{i,j}\mathbf{F}$ and $\alpha \mathbf{A}_i \mathbf{F}_i$. For $\mathbf{Y}$, the space cost of computing $\tilde{\mathbf{R}}_{i,j}$ and $\mathbf{F}_i(\mathbf{F}_j'\mathbf{F}_j)$ is $O(m_{i,j})$ and $O(n_i r)$, respectively. Therefore, the space complexity of calculating $\sum_{j:\, \mathbf{G}(i,j)=1} ((1 - w^2)\tilde{\mathbf{R}}_{i,j} + w^2 \mathbf{F}_i \mathbf{F}_j')\mathbf{F}$ is $O(\max_{j:\, \mathbf{G}(i,j)=1} m_{i,j} + n_i r)$. On the other hand, the space required to compute $\alpha \mathbf{T}_i \mathbf{F}_i + \beta \mathbf{F}_i$

$O(n_i r)$. Putting all together, the space cost of updating all low-rank matrices in each iteration is of $O(\max\limits_{i,j:\ \mathbf{G}(i,j)=1} m_{i,j} + \max_i n_i r)$. Thus, the overall space complexity of Algorithm 3.5 is $O(\sum_{i=1}^g (n_i r + m_i) + \sum\limits_{i,j:\ \mathbf{G}(i,j)=1} m_{i,j})$.   □

Variants

Here, we discuss some variants of the proposed FASCINATE algorithm.

Collective One-Class Collaborative Filtering. By setting $w \in (0,1)$, FASCINATE can be used to address one-class collaborative filtering problem, where implicit dependencies extensively exist between nodes from different layers. Specifically, in two-layered networks, FASCINATE is reduced to *wiZAN-Dual*, a weighting-based, dual-regularized, one-class collaborative filtering algorithm proposed in [139].

Multi-layered Network Clustering. By setting all the entries in the weight matrix $\mathbf{W}_{i,j}$ to 1 in Eq. (3.9), we have the following objective function:

$$\min_{\mathbf{F}_i \geq 0(i=1,\ldots,g)} J = \sum_{i,j:\ \mathbf{G}(i,j)=1} \|\mathbf{D}_{i,j} - \mathbf{F}_i \mathbf{F}_j'\|_F^2 + \alpha \sum_{i=1}^g \mathrm{tr}(\mathbf{F}_i'(\mathbf{T}_i - \mathbf{A}_i)\mathbf{F}_i) + \beta \sum_{i=1}^g \|\mathbf{F}_i\|_F^2,$$
(3.36)

where $\mathbf{F}_i$ can be viewed as the cluster membership matrix for nodes in layer-i. By following similar procedure in FASCINATE, we can get the local optima of the above objective function with the following updating rule:

$$\mathbf{F}_i(u,v) \leftarrow \mathbf{F}_i(u,v) \sqrt{\frac{\mathbf{X}_c(u,v)}{\mathbf{Y}_c(u,v)}},$$
(3.37)

where

$$\mathbf{X}_c = \sum_{j:\ \mathbf{G}(i,j)=1} \mathbf{D}_{i,j}\mathbf{F}_j + \alpha \mathbf{A}_i \mathbf{F}_i$$
(3.38)

$$\mathbf{Y}_c = \sum_{j:\ \mathbf{G}(i,j)=1} \mathbf{F}_i \mathbf{F}_j' \mathbf{F}_j + \alpha \mathbf{T}_i \mathbf{F}_i + \beta \mathbf{F}_i.$$
(3.39)

Although in the above updating rule, we do not need to calculate $\tilde{\mathbf{R}}_{i,j}$ for $\mathbf{Y}_c$ comparing to $\mathbf{Y}$ in Eq. (3.15), the overall time complexity for the algorithm is still $O([\sum_{i=1}^g (\sum_{j:\ \mathbf{G}(i,j)=1} (m_{i,j}r + (n_i + n_j)r^2) + m_i r)]t)$. If we restrict ourselves to two-layered networks (i.e., $g = 2$), the above variant for FASCINATE becomes a dual-regularized co-clustering algorithm [77].

Unconstrained Fascinate. In FASCINATE, we place an non-negative constraint on the latent features $\{\mathbf{F}_i\}_{1=1...g}$ in Eq. (3.9) to pursue good interpretability and efficiency. By discarding the non-negative constraint, we have FASCINATE-UN, an unconstrained variant of FASCINATE, which can be solved with gradient descent method as shown in Algorithm 3.6. It first randomly

---

Algorithm 3.6 The FASCINATE-UN Algorithm

Input: (1) A multi-layered network with (a) layer-layer dependency matrix $\mathbf{G}$, (b) within-layer connectivity matrices $\mathcal{A} = \{\mathbf{A}_1, ..., \mathbf{A}_g\}$, and (c) observed cross-layer node dependency matrices $\mathcal{D} = \{\mathbf{D}_{i,j}\ i, j = 1, ..., g\}$; (2) the rank size r; (3) weight w; (4) regularized parameters α and β; and (5) parameters $a \in (0, 0.5)$, $b \in (0, 1)$

Output: Low-rank representations for each layer $\{\mathbf{F}_i\}_{i=1,...,g}$

1: **for** $i = 1$ to g **do**
2:    initialized $\mathbf{F}_i$ as $n_i \times r$ random matrix
3: **end for**
4: **while** not converge **do**
5:    **for** $i = 1$ to g **do**
6:       compute $\frac{\partial J_i}{\partial \mathbf{F}_i}$ with Eq. (3.11)
7:       $\tau \leftarrow$ step size from backtracking line search
8:       $\mathbf{F}_i \leftarrow \mathbf{F}_i - \tau \frac{\partial J_i}{\partial \mathbf{F}_i}$
9:    **end for**
10: **end while**
11: return $\{\mathbf{F}_i\}_{i=1,...,g}$

---

initializes the low-rank matrices for each layer (steps 1–3) and then begins the iterative update procedure. In each iteration (steps 4–10), the algorithm alternatively updates $\{\mathbf{F}_i\}_{i=1,...,g}$ with gradient descent method one by one. Similar to FASCINATE, the two criteria we use to terminate the iteration are: (1) either the difference of the objective function (J in Eq. (3.9)) between two successive iterations is less than a threshold ξ, or (2) the maximum iteration number t is reached. The complexity of computing $\frac{\partial J_i}{\partial \mathbf{F}_i}$ is the same as the complexity of computing $\mathbf{X}$ and $\mathbf{Y}$ in Algorithm 3.5. However, in the backtracking line search procedure in step 7, calculating the value of the objective function J_i is required to find step size τ with complexity $O(\sum_{j:\ G(i,j)=1} n_i n_j r + n_i^2 r)$. This quadratic complexity would increase the overall complexity of Algorithm 3.6 significantly in large systems.

Collective Matrix Factorization. Instead of exploiting node homophily effect from each layer, we can view the with-in layer networks as additional constraints for matrix factorization problem as modeled in the following objective function

$$\min_{\mathbf{F}_i \geq 0 (i=1,\dots,g)} \sum_{i,j:\ G(i,j)=1} \|\mathbf{W}_{i,j} \odot (\mathbf{D}_{i,j} - \mathbf{F}_i \mathbf{F}_j')\|_F^2 + \alpha \sum_{i=1}^{g} \|\mathbf{A}_i - \mathbf{F}_i \mathbf{F}_i'\|_F^2 + \beta \sum_{i=1}^{g} \|\mathbf{F}_i\|_F^2,$$

$$(3.40)$$

where $\mathbf{F}_i$ is the latent features for nodes in layer-i.

Again, the above problem can be solved with similar procedure in FASCINATE. The updating rules are as follows:

$$\mathbf{F}_i(u,v) \leftarrow \mathbf{F}_i(u,v) \sqrt{\frac{\mathbf{X}_{col}(u,v)}{\mathbf{Y}_{col}(u,v)}},$$

$$(3.41)$$

where $\mathbf{X}_{col}$ and $\mathbf{Y}_{col}$ are defined as

$$\mathbf{X}_{col} = \sum_{j:\ G(i,j)=1} \mathbf{D}_{i,j} \mathbf{F}_j + 2\alpha \mathbf{A}_i \mathbf{F}_i$$

$$(3.42)$$

$$\mathbf{Y}_{col} = \sum_{j:\ G(i,j)=1} ((1-w^2)\tilde{\mathbf{R}}_{i,j} + w^2 \mathbf{F}_i \mathbf{F}_j')\mathbf{F}_j + 2\alpha \mathbf{F}_i \mathbf{F}_i' \mathbf{F}_i + \beta \mathbf{F}_i.$$

$$(3.43)$$

The complexity of the above method is of the same order with FASCINATE. In particular, when the within-layer connectivity matrices $\mathcal{A} = \{\mathbf{A}_1, \dots, \mathbf{A}_g\}$ are absent, the proposed FASCINATE can be viewed as a collective matrix factorization method in [112].

While the proposed FASCINATE includes these existing methods as its special cases, its major advantage lies in its ability to collectively leverage all the available information (e.g., the within-layer connectivity, the observed cross-layer dependency) for dependency inference. As we will demonstrate in the experimental section, such a methodical strategy leads to substantial and consistent inference performance boosting. Nevertheless, a largely unanswered question for these methods (including FASCINATE) is how to handle *zero-start* nodes. That is, when a new node arrives with no observed cross-layer dependencies, how can we effectively and efficiently infer its dependencies without rerunning the algorithm from scratch. In the next section, we present a *sub-linear* algorithm to solve this problem (i.e., Problem 3.4).

3.2.3   PROPOSED ALGORITHM FOR CODE-ZERO

A multi-layered network often exhibits high dynamics, e.g., the arrival of new nodes. For example, for a newly identified chemical in the biological system, we might know how it interacts with some existing chemicals (i.e., the within-layer connectivity). However, its cross-layer dependencies w.r.t. proteins and/or diseases might be completely unknown. This section addresses such *zero-start* problems (i.e., Problem 3.4). Without loss of generality, we assume that the newly added node resides in layer-*1*, indexed as its $(n_1 + 1)$th node. The within-layer connectivity between the newly added node and the existing n_1 nodes is represented by a $1 \times n_1$ row vector $\mathbf{s}$, where $\mathbf{s}(u)$ $(u = 1, ..., n_1)$ denotes the (within-layer) connectivity between the newly added node and the uth existing node in layer-1.

We could just rerun our FASCINATE algorithm on the entire multi-layered network with the newly added node to get its low-rank representation (i.e., a $1 \times r$ row vector $\mathbf{f}$), based on which its cross-layer dependencies can be estimated. However, the running time of this strategy is linear w.r.t. the size of the *entire* multi-layered network. For example, on a three-layered infrastructure network whose size is in the order of 14 million, it would take FASCINATE $2,500+$ s to update the low-rank matrices $\{\mathbf{F}_i\}$ for a *zero-start* node with rank $r = 200$, which might be too costly in online settings. In contrast, our upcoming algorithm is *sub-linear*, and it only takes less than 0.001 s on the same network without jeopardizing the accuracy.

There are two key ideas behind our online algorithm. The first is to view the newly added node as a perturbation to the original network. In detail, the updated within-layer connectivity matrix $\hat{\mathbf{A}}_1$ for layer-1 can be expressed as

$$\hat{\mathbf{A}}_1 = \left[\begin{array}{cc} \mathbf{A}_1 & \mathbf{s}' \\ \mathbf{s} & 0 \end{array} \right], \tag{3.44}$$

where $\mathbf{A}_1$ is the within-layer connectivity matrix for layer-1 before the arrival of the new node.

Correspondingly, the updated low-rank representation matrix for layer-1 can be expressed as $\hat{\mathbf{F}}_1 = [\hat{\mathbf{F}}'_{1(n_1 \times r)} \ \mathbf{f}']'$, where $\hat{\mathbf{F}}_{1(n_1 \times r)}$ is the updated low-rank representation for the existing n_1 nodes in layer-1. Then the new objective function $\hat{J}$ in Eq. (3.9) can be reformatted as

$$\hat{J} = \sum_{\substack{i,j:\ G(i,j)=1 \\ i,j \neq 1}} \|\mathbf{W}_{i,j} \odot (\mathbf{D}_{i,j} - \hat{\mathbf{F}}_i \hat{\mathbf{F}}'_j)\|^2_F + \sum_{j:\ G(1,j)=1} \|\hat{\mathbf{W}}_{1,j} \odot (\hat{\mathbf{D}}_{1,j} - \hat{\mathbf{F}}_1 \hat{\mathbf{F}}'_j)\|^2_F \tag{3.45}$$

$$+ \sum_{i=2}^{g} \frac{\alpha}{2} \sum_{u=1}^{n_i} \sum_{v=1}^{n_i} \mathbf{A}_i(u,v) \|\hat{\mathbf{F}}_i(u,:) - \hat{\mathbf{F}}_i(v,:)\|^2_2$$

$$+ \frac{\alpha}{2} \sum_{u=1}^{n_1} \sum_{v=1}^{n_1} \mathbf{A}_1(u,v) \|\hat{\mathbf{F}}_1(u,:) - \hat{\mathbf{F}}_1(v,:)\|^2_2$$

$$+ \beta \sum_{i=2}^{g} \|\hat{\mathbf{F}}_i\|^2_F + \beta \|\hat{\mathbf{F}}'_{1(n_1 \times r)}\|^2_F + \alpha \sum_{v=1}^{n_1} \mathbf{s}(v) \|\mathbf{f} - \hat{\mathbf{F}}_1(v,:)\|^2_2 + \beta \|\mathbf{f}\|^2_2.$$

Since the newly added node has no dependencies, we can set

$$\hat{\mathbf{W}}_{1,j} = \left[\begin{array}{c} \mathbf{W}_{1,j} \\ \mathbf{0}_{(1 \times n_j)} \end{array} \right], \quad \hat{\mathbf{D}}_{1,j} = \left[\begin{array}{c} \mathbf{D}_{1,j} \\ \mathbf{0}_{(1 \times n_j)} \end{array} \right].$$

Therefore, the second term in $\hat{J}$ can be simplified as

$$\sum_{j:\ G(1,j)=1} \|\mathbf{W}_{1,j} \odot (\mathbf{D}_{1,j} - \hat{\mathbf{F}}_{1(n_1 \times r)} \hat{\mathbf{F}}'_j)\|^2_F. \tag{3.4}$$

Combining Eq. (3.45), Eq. (3.46), and J in Eq. (3.9) together, $\hat{J}$ can be expressed as

$$\hat{J} = J + J^1, \tag{3.47}$$

where $J^1 = \alpha \sum_{v=1}^{n_1} s(v)\|\mathbf{f} - \hat{\mathbf{F}}_1(v,:)\|_2^2 + \beta\|\mathbf{f}\|_2^2$, and J is the objective function without the newly arrived node.

The second key idea of our online algorithm is that in Eq. (3.47), J is often orders of magnitude larger than J^1. For example, in the **BIO** dataset used in Section 3.2.3, J is in the order of 10^3, while J^1 is in the order of 10^{-1}. This naturally leads to the following approximation strategy, that is, we (1) fix J with $\{\mathbf{F}_i^*\}_{i=1,\ldots,g}$ (i.e., the previous local optimal solution to Eq. (3.9) without the newly arrived node), and (2) optimize J^1 to find out the low-rank representation $\mathbf{f}$ for the newly arrived node. That is, we seek to solve the following optimization problem:

$$\mathbf{f} = \arg\min_{\mathbf{f} \geq 0} J^1 \quad \text{subject to: } \hat{\mathbf{F}}_{1(n_1 \times r)} = \mathbf{F}_1^* \tag{3.48}$$

with which we can get an approximate solution $\{\hat{\mathbf{F}}_i\}_{i=1,\ldots,g}$ to $\hat{J}$.

To solve $\mathbf{f}$, we take the derivative of J^1 w.r.t. $\mathbf{f}$ and get

$$\frac{1}{2}\frac{\partial J^1}{\partial \mathbf{f}} = \beta\mathbf{f} + \alpha \sum_{v=1}^{n_1} s(v)(\mathbf{f} - \mathbf{F}_1^*(v,:)) \tag{3.49}$$

$$= \left(\beta + \alpha \sum_{v=1}^{n_1} s(v)\right)\mathbf{f} - \alpha\mathbf{s}\mathbf{F}_1^*.$$

Since α and β are positive, the Hessian matrix of J^1 is a positive diagonal matrix. Therefore, the global minimum of J^1 can be obtained by setting its derivative to zero. Then the optimal solution to J^1 can be expressed as

$$\mathbf{f} = \frac{\alpha\mathbf{s}\mathbf{F}_1^*}{\beta + \alpha \sum_{v=1}^{n_1} s(v)}. \tag{3.50}$$

For the newly added node, $\mathbf{f}$ can be viewed as the weighted average of its neighbors' low-rank representations. Notice that in Eq. (3.50), the non-negativity constraint on $\mathbf{f}$ naturally holds. Therefore, we refer to this solution (i.e., Eq. (3.50)) as FASCINATE-ZERO. In this way, we can successfully decouple the cross-layer dependency inference problem for the *zero-start* node from the entire multi-layered network and localize it only among its neighbors in layer-1. The localization significantly reduces the time complexity, as summarized in Lemma 3.10, which is linear w.r.t. the number of neighbors of the new node (and therefore is *sub linear* w.r.t. the size of the entire network).

Lemma 3.10 *Let nnz(**s**) denotes the total number of within-layer links between the newly added node and the original nodes in layer-1 (i.e., nnz(**s**) is the degree for the newly added node). Then the time complexity of FASCINATE-ZERO is $O(nnz(\mathbf{s})r)$.*

Proof. Since the links between the newly added node and the original nodes in layer-*1* are often very sparse, the number of non-zero elements in $\mathbf{s}$ (nnz($\mathbf{s}$)) is much smaller than n_1. Therefore, the complexity of computing $\mathbf{sF}_1^*$ can be reduced to $O(\text{nnz}(\mathbf{s})r)$. The multiplication between α and $\mathbf{sF}_1^*$ takes $O(r)$. Computing $\sum_{v=1}^{n_1} \mathbf{s}(v)$ takes $O(\text{nnz}(\mathbf{s}))$. Thus, the overall complexity of computing $\mathbf{f}$ is $O(\text{nnz}(\mathbf{s})r)$. $\square$

Remarks. Following the similar procedure in FASCINATE-ZERO, it is easy to extend the zero-start problem to the scenario where a new within-layer edge is added to two existing nodes. Suppose that in layer-*1* a new edge $\langle u, v \rangle$ is added between node u and node v. To find out the updated low-rank matrices $\{\hat{\mathbf{F}}_i\}$ efficiently after the perturbation, we can partition the nodes in the multi-layered network into two parts: (1) nodes that can be affected by either node u or node v (denoted as $\mathcal{N}^{\{u,v\}}$) and (2) nodes that are irrelevant to both node u and node v (denoted as $\mathcal{N}^{\backslash\{u,v\}}$). Specifically, we define that node w can be affected by node u if and only if there exists a path from u to w, and the links in the path can be either within-layer edges or cross-layer dependencies; otherwise, node w is viewed as irrelevant to u. By this definition, we have $\mathcal{N}^{\{u,v\}} \cap \mathcal{N}^{\backslash\{u,v\}} = \Phi$ and the new objective function $\hat{J}$ can be decomposed into two parts as

$$\hat{J} = \hat{J}^{\{u,v\}} + \hat{J}^{\backslash\{u,v\}}, \tag{3.51}$$

where $\hat{J}^{\{u,v\}}$ only contains the optimization terms for the latent features of the affected nodes ($\{\hat{\mathbf{F}}_i\}^{\{u,v\}}$), while $\hat{J}^{\backslash\{u,v\}}$ contains the terms for latent features of irrelevant nodes ($\{\hat{\mathbf{F}}_i\}^{\backslash\{u,v\}}$). As the newly added edge $\langle u, v \rangle$ in layer-*1* would not cause any changes in $\hat{J}^{\backslash\{u,v\}}$, $\{\hat{\mathbf{F}}_i\}^{\backslash\{u,v\}}$ would remain the same with the previous local optima solution $\{\mathbf{F}_i^*\}^{\backslash\{u,v\}}$. Therefore, the only terms we need to optimize is $\hat{J}^{\{u,v\}}$ w.r.t. the affected latent features $\{\hat{\mathbf{F}}_i\}^{\{u,v\}}$.

3.2.4   EXPERIMENTAL EVALUATION

In this section, we evaluate the proposed FASCINATE algorithms. All experiments are designed to answer the following questions.

- *Effectiveness.* How effective are the proposed FASCINATE algorithms in inferring the missing cross-layer dependencies?

- *Efficiency.* How fast and scalable are the proposed algorithms?

Experimental Setup

Datasets Description. We perform our evaluations on five different datasets, including (1) a three-layer, cross-domain, paper citation network in the academic research domain (CITATION); (2) a five-layer Italy network in the critical infrastructure domain (INFRA-5); (3) a three-layer network in the critical infrastructure domain (INFRA-3); (4) a three-layer CTD (Comparative Toxicogenomics Database) network in the biological domain (BIO); and (5) a three-layer Aminer academic network in the social collaboration domain (SOCIAL). Th

Table 3.3: Statistics of datasets

Dataset	Number of Layers	Number of Nodes	Number of Links	Number of Crosslinks
CITATION	3	33,249	27,017	4,589
INFRA-5	5	349	379	565
INFRA-3	3	15,126	29,861	28,023,500
SOCIAL	3	125,344	214,181	188,844
BIO	3	35,631	253,827	75,456

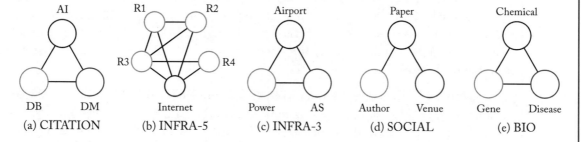

Figure 3.18: The abstract dependency structure of each dataset.

statistics of these datasets are shown in Table 3.3, and the abstract layer-layer dependency graphs of these four datasets are summarized in Figure 3.18. In all these four datasets, the cross-layer dependencies are binary and undirected (i.e., $\mathbf{D}_{i,j}(u,v) = \mathbf{D}_{j,i}(v,u)$).

CITATION. The construction of this publication network is based on the work in [72]. It contains three layers, which correspond to the paper citation networks in AI (Artificial Intelligence), DB (Database), and DM (Data Mining) domains. The cross-domain citations naturally form the cross-layer dependencies in the system. For example, the cross-layer dependency between the AI layer and the DM layer indicates the citations between AI papers and DM papers. The papers in the system are from the top conferences in the corresponding areas as shown in Table 3.4. The number of nodes in each layer varies from 5,158–18,243, and the number of within layer links ranges from 20,611–40,885. The number of cross layer dependencies ranges from 536–2,250. The structure of the entire system is shown in Figure 3.18a.

INFRA-5. The construction of this critical infrastructure network is based on the data implicated from an electrical blackout in Italy in September 2003 [102]. It contains five layers, including four layers of regional power grids and one Internet network [102]. The regional power

Table 3.4: List of conferences in each domain

Domain	AI	DM	DB
Conferences	IJCAI	KDD	SIGMOD
	AAAI	ICDM	VLDB
	ICML	SDM	ICDM
	NIPS	PKDD	PODS

grids are partitioned by macroregions.[3] To make the regional networks more balanced, we merge the Southern Italy power grid and the Island power grid together. The power transfer lines between the four regions are viewed as cross-layer dependencies. For the Italy Internet network, it is assumed that each Internet center is supported by the power stations within a radius of 70 km. Its abstract dependency graph is shown in Figure 3.18b. The smallest layer in the network has 39 nodes and 50 links, while the largest network contains 151 nodes and 158 links. The number of dependencies is up to 307.

INFRA-3. This dataset contains the following three critical infrastructure networks: an airport network,[4] an autonomous system network,[5] and a power grid [133]. We construct a three-layered network in the same way as [15]. The three infrastructure networks are functionally dependent on each other. Therefore, they form a triangle-shaped multi-layered network as shown in Figure 3.18c. The construction of the cross-layer dependencies is based on geographic proximity.

SOCIAL. This dataset contains three layers, including a collaboration network among authors, a citation network between papers, and a venue network [117]. The number of nodes in each layer ranges from 899–62,602, and the number of within-layer links ranges from 2,407–201,037. The abstract layer-layer dependency graph of SOCIAL is shown in Figure 3.18d. The collaboration layer is connected to the paper layer with the authorship dependency, while the venue layer is connected to the paper layer with publishing dependency. For the *Paper-Author* dependency, we have 126,242 links across the two layers; for the *Paper-Venue* dependency, we have 62,602 links.

BIO. The construction of the CTD network is based on the works in [30, 100, 126]. It contains three layers, which are chemical, disease, and gene similarity networks. The number of nodes in these networks ranges from 4,256–25,349, and the number of within-layer links ranges from 30,551–154,167. The interactions between chemicals, genes, and diseases form the cross-layer dependency network as shown in Figure 3.18e. For *Chemical-Gene* dependency, we

---

[3]https://en.wikipedia.org/wiki/First-level_NUTS_of_the_European_Union
[4]http://www.levmuchnik.net/Content/Networks/NetworkData.html
[5]http://snap.stanford.edu/data/

have 53,735 links across the two layers; for *Chemical-Disease* dependency, we have 19,771 links; and for *Gene-Disease* dependency, we have 1,950 links.

For all datasets, we randomly select 50% cross-layer dependencies as the training set and use the remaining 50% as the test set.

Comparing Methods. We compare FASCINATE with the following methods: (1) FASCINATE-CLUST—a variant of the proposed method for the purpose of dependency clustering; (2) FASCINATE-UN—a variant of FASCINATE without non-negative constrain; (3) *MulCol*—a collective matrix factorization method [112]; (4) *PairSid*—a pairwise one-class collaborative filtering method proposed in [139]; (5) *PairCol*—a pairwise collective matrix factorization method degenerated from *MulCol*; (6) *PairNMF*—a pairwise non-negative matrix factorization (*NMF*) based method [76]; (7) *PairRec*—a pairwise matrix factorization based algorithm introduced in [59]; (8) *FlatNMF*—an NMF-based method that treats the input multi-layered network as a flat-structured single network (i.e., by putting the within-layer connectivity matrices in the diagonal blocks, and the cross-layer dependency matrices in the off-diagonal blocks); and (9) *FlatRec*—a matrix factorization-based method using the same techniques as *PairRec* but treating the input multi-layered network as a single network as in *FlatNMF*.

For the experimental results reported in this work, we set rank $r = 100$, maximum iteration $t = 100$, termination threshold $\xi = 10^{-8}$, weight $w^2 = 0.1$, regularization parameters $\alpha = 0.1$, $\beta = 0.1$, and backtracking line search parameters $a = 0.1$, $b = 0.8$ unless otherwise stated.

Evaluation Metrics. We use the following metrics for the effectiveness evaluations.

- **MAP**. It measures the mean average precision over all entities in the cross-layer dependency matrices [75]. A larger *MAP* indicates better inference performance.

- **R-MPR**. It is a variant of Mean Percentage Ranking for one-class collaborative filtering [47]. *MPR* is originally used to measure the user's satisfaction with items in a ranked list. In our case, we can view the nodes from one layer as users, and the nodes of the dependent layer(s) as items. The ranked list therefore can be viewed as ordered dependencies by their importance. Smaller *MPR* indicates better inference performance. Specifically, for a randomly produced list, its *MPR* is expected to be 50%. Here, we define R-MPR = 0.5 − MPR, so that larger *R-MRP* indicates better inference performance.

- **HLU**. Half-Life Utility is also a metric from one-class collaborative filtering. By assuming that the user will view each consecutive item in the list with exponential decay of possibility, it estimates how likely a user will choose an item from a ranked list [95]. In our case, it measures how likely a node will establish dependencies with the nodes in the ranked list. A larger *HLU* indicates better inference performance.

- **AUC**. Area Under ROC Curve is a metric that measures classification accuracy. A larger *AUC* indicates better inference performance.

- **Prec@K**. Precision at K is defined by the proportion of true dependencies among the top K inferred dependencies. A larger *Prec@K* indicates better inference performance.

Machine and Repeatability All the experiments are performed on a machine with two processors Intel Xeon 3.5 GHz with 256 GB of RAM. The algorithms are programmed with MAT-LAB using a single thread.

Effectiveness

In this section, we aim to answer the following three questions: (1) How effective is FASCINATE for Problem 3.1 (i.e., CODE)? (2) How effective is FASCINATE-ZERO for Problem 3.4 (i.e., CODE-ZERO)? (3) How sensitive are the proposed algorithms w.r.t. the model parameters?

Effectiveness of Fascinate. We compare the proposed algorithms and the existing methods on all five datasets. As FASCINATE-UN is not scalable to large networks, we only evaluate its performance on two small datasets—CITATION and INFRA-5. The results are shown in Table 3.5 and Table 3.6. There are several interesting observations. First is that our proposed FASCINATE algorithm and its variants (FASCINATE-CLUST and FASCINATE-UN) consistently outperform all other methods in terms of all the five evaluation metrics. Second, by exploiting the structure of the multi-layered network, FASCINATE, FASCINATE-CLUST, FASCINATE-UN, and *MulCol* can achieve significantly better performance than the pairwise methods in most datasets. Third, among the pairwise baselines, *PairSid* and *PairCol* are better than *PairNMF* and *PairRec*. The main reason is that the first two algorithms utilize both within-layer connectivity matrices and cross-layer dependency matrix for matrix factorization, while the latter two only use the observed dependency matrix. Finally, the relatively poor performance of *FlatNMF* and *FlatRec* implies that simply flattening the multi-layered network into a single network is insufficient to capture the intrinsic correlations across different layers.

We also test the sensitivity of the proposed algorithms w.r.t. the sparsity of the observed cross-layer dependency matrices (i.e., the ratio of the missing values) on INFRA-3. The results in Figure 3.19 demonstrate that both FASCINATE and FASCINATE-CLUST perform well even when 90%+ entries in the dependency matrices are missing.

Effectiveness of Fascinate-ZERO. To evaluate the effectiveness of FASCINATE-ZERO, we randomly select one node from the *Chemical* layer in the BIO dataset as the newly arrived node and compare the inference performance between FASCINATE-ZERO and FASCINATE. The average results over multiple runs are presented in Figure 3.20. We can see that FASCINATE-ZERO bears a very similar inference power as FASCINATE, but it is orders of magnitude faster. We observe similar performance when the *zero-start* nodes are selected from the other two layers (i.e., *Gene* and *Disease*).

Parameter Studies. There are three parameters—α, β, and r—in the proposed FASCINATE algorithm. α is used to control the impact of node homophily, β is used to avoid over-fitting, and

Table 3.5: Cross-layer dependency inference on CITATION

Methods	MAP	R-MPR	HLU	AUC	Prec@10
Fascinate	0.1389	**0.3907**	19.1264	**0.8523**	0.0428
Fascinate-Clust	0.1347	0.3882	19.8367	0.8487	0.0407
Fascinate-UN	**0.1873**	0.2685	**25.1961**	0.7423	**0.0532**
MulCol	0.1347	0.3882	19.8367	0.8487	0.0459
PairSid	0.1623	0.3868	21.8641	0.8438	0.0480
PairCol	0.1311	0.3838	19.1697	0.8388	0.0446
PairNMF	0.0338	0.1842	4.4397	0.6009	0.0103
PairRec	0.0351	0.2582	5.3407	0.6527	0.0129
FlatNMF	0.0811	0.3539	12.1835	0.8084	0.0284
FlatRec	0.0032	0.3398	0.0608	0.8113	0.0001

Table 3.6: Cross-layer dependency inference on INFRA-5

Methods	MAP	R-MPR	HLU	AUC	Prec@10
Fascinate	**0.5040**	**0.3777**	**67.2231**	**0.8916**	**0.2500**
Fascinate-Clust	0.4297	0.3220	56.8215	0.8159	0.2340
Fascinate-UN	0.4354	0.3631	60.2393	0.8575	0.2412
MulCol	0.4523	0.3239	59.8115	0.8329	0.2413
PairSid	0.3948	0.2392	49.5484	0.7413	0.2225
PairCol	0.3682	0.2489	48.5966	0.7406	0.2309
PairNMF	0.1315	0.0464	15.7148	0.5385	0.0711
PairRec	0.0970	0.0099	9.4853	0.5184	0.0399
FlatNMF	0.3212	0.2697	44.4654	0.7622	0.1999
FlatRec	0.1020	0.0778	11.5598	0.5740	0.0488

r is the number of columns of the low-rank matrices $\{\mathbf{F}_i\}$. We fix one of these parameters and study the impact of the remaining two on the inference results. From Figure 3.21, we can see that MAP is stable over a wide range of both α and β. As for the third parameter r, the inference performance quickly increases w.r.t. r until it hits 200, after which the MAP is almost flat. This suggests that relatively small size of the low-rank matrices might be sufficient to achieve a satisfactory inference performance.

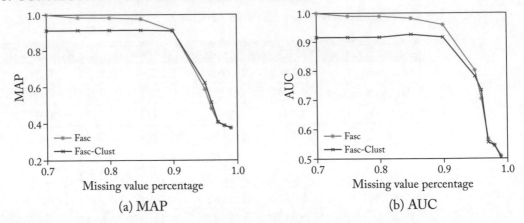

Figure 3.19: Performance of FASCINATE and FASCINATE-CLUST on INFRA-3 dataset under different missing value percentages.

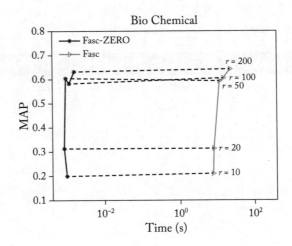

Figure 3.20: The effectiveness of FASCINATE-ZERO in BIO network w.r.t. different rank r.

For FASCINATE-UN, we study the impact of the backtracking line search parameters on its performance. By fixing α, β, and rank r to 0.1, 0.1, and 100, respectively, we examine wide range of a and b within their domains as shown in Figure 3.22. We can see that the inference performance is sensitive to the combination of a and b because subtle parameter changes may affect the convergence speed in Algorithm 3.6 greatly, which would have an impact on the inference performance within limited iterations consequently.

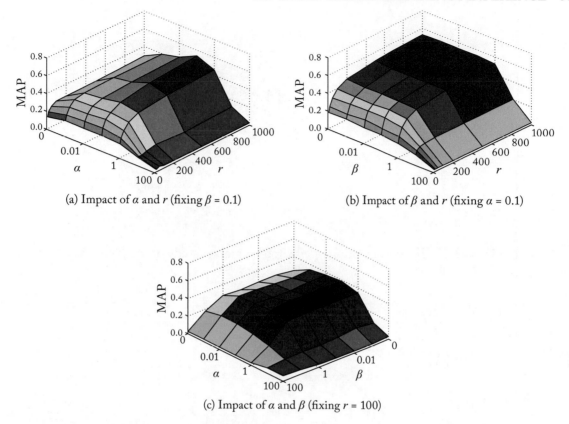

(a) Impact of α and r (fixing $\beta = 0.1$)

(b) Impact of β and r (fixing $\alpha = 0.1$)

(c) Impact of α and β (fixing $r = 100$)

Figure 3.21: The parameter studies of the Bio dataset.

Efficiency

The scalability results of FASCINATE and FASCINATE-ZERO are presented in Figure 3.23. As we can see in Figure 3.23a, FASCINATE scales linearly w.r.t. the overall network size (i.e., $\sum_i (n_i + m_i) + \sum_{i,j} m_{i,j}$), which is consistent with our previous analysis in Lemma 3.8. As for FASCINATE-ZERO, it scales *sub-linearly* w.r.t. the entire network size. This is because, by Lemma 3.10, the running time of FASCINATE-ZERO is only dependent on the neighborhood size of the newly added node, rather than that of the entire network. Finally, we can see that FASCINATE-ZERO is much more efficient than FASCINATE. To be specific, on the entire IN-FRA-3 dataset, FASCINATE-ZERO is 10,000,000+ faster than FASCINATE (i.e., 1.878×10^{-4} s vs. 2.794×10^3 s).

In addition, we compare the running time of FASCINATE and FASCINATE-UN on CITA-TION and INFRA-5 networks. The results are as shown in Figure 3.24. As we can see, FAS-

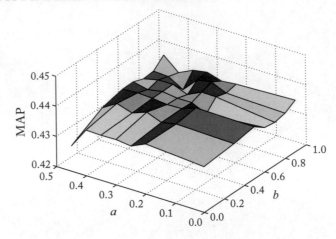

Figure 3.22: The backtracking line search parameter study of the INFRA-5 dataset.

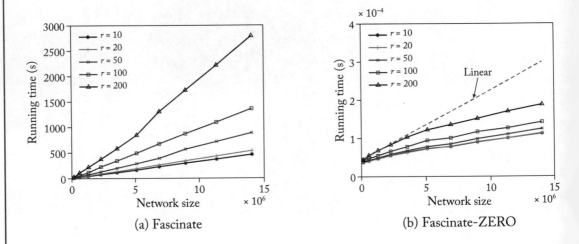

Figure 3.23: Wall-clock time vs. the size of the network.

cinate-UN is orders of magnitude slower than FASCINATE to achieve similar inference results which is consistent with our complexity analysis.

3.3 INCREMENTAL ONE-CLASS COLLABORATIVE FILTERING

The past decade has witnessed the prosperity of recommender systems in various application ranging from e-commerce platforms to online service providers. Among the numerous recon

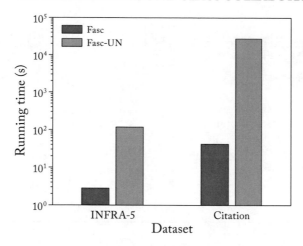

Figure 3.24: Wall-clock running time of FASCINATE and FASCINATE-UN.

mendation algorithms in the literature, collaborative filtering-based methods are widely adopted in many applications due to their superior effectiveness. Traditional collaborative filtering algorithms are typically designed to provide recommendations based on users' explicit, multi-scale feedback (e.g., rating 1–5). However, in many real applications, the preferences might only be inferred from users' implicit, one-class feedback (e.g., actions or inactions). For example, it is reasonable to infer that a user likes a song if s/he listened to it from the beginning to the end; otherwise, s/he may not be into the song. Such applications are generally formulated as one-class collaborative filtering (OCCF) problems [95].

The key challenges for OCCF lie in the sparsity of positive feedback (preferences) and the ambiguity of missing preferences. A promising way to address those issues is to exploit side information from the social networks of users and/or similarity networks of items as in [79] and [139]. The key idea behind those methods is that socially connected users tend to share similar tastes on items; while similar items are more likely to impose similar impact on users.

Most of the existing OCCF algorithms are focused on *static* systems, despite the fact that both user preferences and side networks are evolving over time. For example, in the e-commerce platform shown in Figure 3.25, new friendship relations (black dashed lines) and user preferences (red dashed lines) are emerging over time. Meanwhile, as new products are being released to the market, similarity links between newly released products and existing products would appear in the system as well. In such a coupled system, the emergence of new connections and preferences may cause a ripple effect on the platform, hence affecting the preferences of a large proportion of users. Consequently, it is necessary to update the latent features of users and items obtained in the previous time stamp to accommodate changes. A straightforward way to update the latent features is to rerun the OCCF algorithm from scratch whenever the system changes. However,

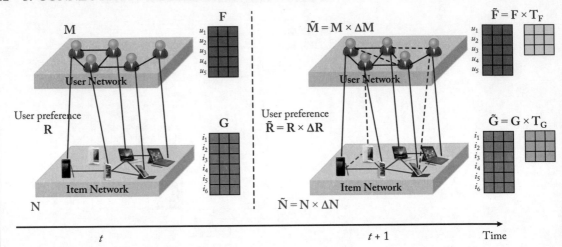

Figure 3.25: An illustration of online one-class recommendation problem with side networks. Solid lines represent the links in the original system, dashed lines represent the newly emerged links. (Best viewed in color.)

for large-scale applications, such a strategy may take an unaffordable long time, which would compromise user experience.

In this work, we propose an efficient algorithm to incrementally update one-class collaborative filtering results with co-evolving side networks. To efficiently accommodate the changes in the system, we propose to model the evolution of latent features based on the following observations: the system often evolves smoothly between two consecutive time stamps such that a large number of observed preferences and network links remain unchanged. Thus, we can view the new latent features as a subtle linear transformation from the previous features. This would in turn allow us to incrementally solve the OCCF problem in a timely manner without re-solving it from scratch.

The main contributions of this work can be summarized as follows.

- *Problem Formulation.* We formally define the problem of incremental OCCF with co-evolving side networks.

- *Algorithms and Analysis.* We propose an incremental OCCF algorithm (i.e., ENCORE) that can efficiently accommodate system dynamics, and analyze its optimality and complexity.

- *Evaluations.* We empirically evaluate the proposed method on real-world datasets to verify its effectiveness and efficiency.

Table 3.7: Cross-layer dependency inference on INFRA-3

Methods	MAP	R-MPR	HLU	AUC	Prec@10
Fascinate	0.4780	0.0788	**55.7289**	0.6970	**0.5560**
Fascinate-Clust	**0.5030**	**0.0850**	49.1223	**0.7122**	0.4917
Fascinate-UN	–	–	–	–	–
MulCol	0.4606	0.0641	49.3585	0.6706	0.4930
PairSid	0.4253	0.0526	47.7284	0.5980	0.4773
PairCol	0.4279	0.0528	48.1314	0.5880	0.4816
PairNMF	0.4275	0.0511	48.8478	0.5579	0.4882
PairRec	0.3823	0.0191	38.9226	0.5756	0.3895
FlatNMF	0.4326	0.0594	45.0090	0.6333	0.4498
FlatRec	0.3804	0.0175	38.0550	0.5740	0.3805

Table 3.8: Cross-layer dependency inference on SOCIAL

Methods	MAP	R-MPR	HLU	AUC	Prec@10
Fascinate	0.0660	**0.2651**	**8.4556**	**0.7529**	**0.0118**
Fascinate-Clust	**0.0667**	0.2462	8.2160	0.7351	0.0108
Fascinate-UN	–	–	–	–	–
MulCol	0.0465	0.2450	6.0024	0.7336	0.0087
PairSid	0.0308	0.1729	3.8950	0.6520	0.0062
PairCol	0.0303	0.1586	3.7857	0.6406	0.0056
PairNMF	0.0053	0.0290	0.5541	0.4998	0.0007
PairRec	0.0056	0.0435	0.5775	0.5179	0.0007
FlatNMF	0.0050	0.0125	0.4807	0.5007	0.0007
FlatRec	0.0063	0.1009	0.6276	0.5829	0.0009

3.3.1   PROBLEM DEFINITION

In this section, we first give a formal definition of the studied problem of incremental OCCF with co-evolving side networks. After that, we provide the preliminaries to facilitate the understanding of the proposed algorithm.

The main symbols used in this work are summarized in Table 3.10. We use bold uppercase for matrices (e.g., $\mathbf{A}$) and $\Delta\mathbf{A}$ for the perturbation matrix of $\mathbf{A}$. ˜ sign denotes the notations

Table 3.9: Cross-layer dependency inference on BIO

Methods	MAP	R-MPR	HLU	AUC	Prec@10
Fascinate	0.3979	0.4066	45.1001	0.9369	0.1039
Fascinate-Clust	0.3189	0.3898	37.4089	0.9176	0.0857
Fascinate-UN	–	–	–	–	–
MulCol	0.3676	0.3954	42.8687	0.9286	0.0986
PairSid	0.3623	0.3403	40.4048	0.8682	0.0941
PairCol	0.3493	0.3153	38.4364	0.8462	0.0889
PairNMF	0.1154	0.1963	15.8486	0.6865	0.0393
PairRec	0.0290	0.2330	3.6179	0.7105	0.0118
FlatNMF	0.2245	0.2900	26.1010	0.8475	0.0615
FlatRec	0.0613	0.3112	8.4858	0.8759	0.0254

after adding the perturbations into the system (i.e., $\tilde{\mathbf{A}} = \mathbf{A} + \Delta\mathbf{A}$). $'$ sign denotes the matrix transpose.

With the notations in Table 3.10, we first define the static OCCF problem with side networks as follows.

Definition 3.11   The problem of static OCCF problem with side networks.

Given: $\Gamma =< \mathbf{M}, \mathbf{N}, \mathbf{R} >$ where $\mathbf{M}$ is an $n_u \times n_u$ social network between users; $\mathbf{N}$ is an $n_i \times n_i$ similarity network between items; and $\mathbf{R}$ is an $n_u \times n_i$ user preference matrix, in which $\mathbf{R}(i, j) = 1$ if user i shows preference on item j, otherwise $\mathbf{R}(i, j) = 0$.

Output: The inferred preference between user u and item i.

Based on the above definition, we give the formal definition of incremental OCCF problem with co-evolving side networks.

Problem 3.12   The problem of incremental OCCF with co-evolving side networks.

Given: (1) The original system $\Gamma =< \mathbf{M}, \mathbf{N}, \mathbf{R} >$; (2) the perturbation of the system $\Delta\Gamma = \Delta\mathbf{M}, \Delta\mathbf{N}, \Delta\mathbf{R} >$; (3) the $n_u \times r$ latent feature matrix $\mathbf{F}$ for users in the original system Γ; and (4) the $n_i \times r$ latent feature matrix $\mathbf{G}$ for items in the original system Γ.

Output: The inferred preference between user u and item i in $\tilde{\Gamma} =< \tilde{\mathbf{M}}, \tilde{\mathbf{N}}, \tilde{\mathbf{R}} >$.

Table 3.10: Main symbols

Symbol	Definition and Description
$\mathbf{A}, \mathbf{B}$	adjacency matrices
$\Delta\mathbf{A}$	perturbation matrix of $\mathbf{A}$
$\tilde{\mathbf{A}}$	updated matrix of $\mathbf{A}$
$\mathbf{A}(i, j)$	the element at ith row jth column in $\mathbf{A}$
$\mathbf{A}'$	transpose of matrix $\mathbf{A}$
$\mathbf{M}$	the adjacency matrix of user network
$\mathbf{N}$	the adjacency matrix of item network
$\mathbf{D}_M, \mathbf{D}_N$	the diagonal degree matrices for $\mathbf{M}$ and $\mathbf{N}$
$\mathbf{R}$	the preference matrix for users w.r.t. items
Γ	the recommendation problem with side networks $\Gamma = <\mathbf{M}, \mathbf{N}, \mathbf{R}>$
$\mathbf{W}$	the weight matrix for $\mathbf{R}$
$\mathbf{F}$	the latent feature matrix for users
$\mathbf{G}$	the latent feature matrix for items
n_u, n_i	number of users and items
m_u, m_i	number of edges in $\mathbf{M}$ and $\mathbf{N}$
m_r	number of observed links in $\mathbf{R}$
r	the rank for $\mathbf{F}$ and $\mathbf{G}$
t	the number of iterations

Preliminaries

Under static settings, the OCCF problem with side networks can be solved with the following optimization problem [139]:

$$\min_{\mathbf{F},\mathbf{G}\geq 0} \underbrace{\|\mathbf{W} \odot (\mathbf{R} - \mathbf{F}\mathbf{G}')\|_F^2}_{\text{Matching Observed Ratings}} + \underbrace{\beta(\|\mathbf{F}\|_F^2 + \|\mathbf{G}\|_F^2)}_{\text{Regularization}} \quad (3.52)$$

$$+ \underbrace{\alpha(\text{tr}(\mathbf{F}'(\mathbf{D}_M - \mathbf{M})\mathbf{F}) + \text{tr}(\mathbf{G}'(\mathbf{D}_N - \mathbf{N})\mathbf{G}))}_{\text{Node Homophily}},$$

where $\odot$ is the Hadamard product with $[\mathbf{A} \odot \mathbf{B}](i, j) = \mathbf{A}(i, j)\mathbf{B}(i, j)$. In the above objective function, $\mathbf{R}$ is the user preference matrix; $\mathbf{F}$ and $\mathbf{G}$ are the low-rank latent feature matrices for users and items, respectively; $\mathbf{D}_M$ and $\mathbf{D}_N$ are the diagonal degree matrices for user network $\mathbf{M}$ and item network $\mathbf{N}$ (i.e., $\mathbf{D}_M(u, u) = \sum_k^{n_u} \mathbf{A}(u, k)$, $\mathbf{D}_N(i, i) = \sum_k^{n_i} \mathbf{A}(i, k)$), respectively. $\mathbf{W}$

is an $n_u \times n_i$ weighting matrix, in which $\mathbf{W}(i, j) = 1$ if $\mathbf{R}(i, j) = 1$ (i.e., positive preference observed between user i and item j), otherwise $\mathbf{W}(i, j) \in (0, 1)$ if $\mathbf{R}(i, j) = 0$ (i.e., no preference observed between user i and item j). It is worth mentioning that the weight for unobserved links is used to mitigate its uncertainty between potential positive preferences and negative examples. Consequently, different weighting strategies can be applied in different scenarios. In this work, we set the weight of all unobserved entries to a global value w for the ease of computation.

In Eq. (3.52), the first term is used to match the preferences in matrix $\mathbf{R}$; the second term is to prevent overfitting of the model; and the third term is used to exploit node homophily in side networks. The intuition behind this term is that similar users would hold similar preferences to items (i.e., small $\|\mathbf{F}(u, :) - \mathbf{F}(v, :)\|_2^2$). Correspondingly, similar items would possess similar attractiveness to users (i.e., small $\|\mathbf{G}(i, :) - \mathbf{G}(j, :)\|_2^2$). The entire optimization problem in Eq. (3.52) can be solved by non-negative matrix factorization techniques [63] with time complexity $O(((m_u + m_i + m_r)r + (n_u + n_i)r^2)t)$ (t is the number of iterations). The inferred preference between user u and item i can be estimated by $\mathbf{F}(u, :)\mathbf{G}(i, :)'$, where $\mathbf{F}$ and $\mathbf{G}$ are the local optimal solutions of Eq. (3.52).

3.3.2 PROPOSED ALGORITHM

In this section, we first introduce the proposed algorithm for incremental OCCF with co-evolving side networks. Then we analyze its effectiveness and efficiency.

The Proposed Algorithm

Given a static recommendation input $\Gamma =< \mathbf{M}, \mathbf{N}, \mathbf{R} >$, we can find its low-rank feature matrices $\mathbf{F}$ and $\mathbf{G}$ by solving Eq. (3.52) as shown in the previous section. However, in real applications, networks are evolving over time with perturbation $\Delta\Gamma =< \Delta\mathbf{M}, \Delta\mathbf{N}, \Delta\mathbf{R} >$ from the previous time stamp. Consequently, the low-rank feature matrices should be updated accordingly to provide a more accurate preference estimation. In real applications, systems are often changing smoothly, hence we can assume that the updated user features $\tilde{\mathbf{F}}$ and item features $\tilde{\mathbf{G}}$ still reside in the same feature space with $\mathbf{F}$ and $\mathbf{G}$, but are subtly transformed by the system perturbations. In this way, the updated feature matrices $\tilde{\mathbf{F}}$ and $\tilde{\mathbf{G}}$ can be viewed as a linear transformation from $\mathbf{F}$ and $\mathbf{G}$ as shown in the example in Figure 3.25 (i.e., $\tilde{\mathbf{F}} = \mathbf{FT}_F$, $\tilde{\mathbf{G}} = \mathbf{GT}_G$). Therefore Problem 3.12 is equivalent to finding the transformation matrices $\mathbf{T}_F$ and $\mathbf{T}_G$ for the new time stamp. Hence, the new objective function under perturbation $\Delta\Gamma$ can be written as

$$\min_{\mathbf{T}_F, \mathbf{T}_G} \|\tilde{\mathbf{W}} \odot (\tilde{\mathbf{R}} - \mathbf{FT}_F \mathbf{T}_G' \mathbf{G}')\|_F^2 \qquad (3.53)$$

$$+ \alpha(\mathrm{tr}\mathbf{T}_F' \mathbf{F}'(\mathbf{D}_{\tilde{M}} - \tilde{\mathbf{M}})\mathbf{FT}_F)$$

$$+ \alpha\mathrm{tr}(\mathbf{T}_G' \mathbf{G}'(\mathbf{D}_{\tilde{N}} - \tilde{\mathbf{N}})\mathbf{GT}_G) + \beta(\|\mathbf{FT}_F\|_F^2 + \|\mathbf{GT}_G\|_F^2)$$

$$\mathrm{s.t.} \quad \mathbf{FT}_F, \mathbf{GT}_G \geq 0.$$

Notice that the above objective function imposes a linear constraint on $\mathbf{T}_F$ and $\mathbf{T}_G$ in $\mathbf{FT}_F, \mathbf{GT}_G \geq 0$, which would inevitably increase the computational complexity. We propose to simplify the constraint by replacing it with a non-negative constraint on $\mathbf{T}_F$ and $\mathbf{T}_G$. As $\mathbf{F}$ and $\mathbf{G}$ are non-negative in the first place, their non-negative linear combinations $\mathbf{FT}_F$ and $\mathbf{GT}_G$ are guaranteed to be non-negative as well. Therefore, we can rewrite the above objective function as follows:

$$\min_{\mathbf{T}_F, \mathbf{T}_G \geq 0} \|\tilde{\mathbf{W}} \odot (\tilde{\mathbf{R}} - \mathbf{FT}_F \mathbf{T}'_G \mathbf{G}')\|_F^2 \tag{3.54}$$

$$+ \alpha \operatorname{tr}(\mathbf{T}'_F \mathbf{F}'(\mathbf{D}_{\tilde{M}} - \tilde{\mathbf{M}})\mathbf{FT}_F)$$

$$+ \alpha \operatorname{tr}(\mathbf{T}'_G \mathbf{G}'(\mathbf{D}_{\tilde{N}} - \tilde{\mathbf{N}})\mathbf{GT}_G) + \beta(\|\mathbf{T}_F\|_F^2 + \|\mathbf{T}_G\|_F^2).$$

As the objective function in Eq. (3.54) is not jointly convex w.r.t. $\mathbf{T}_F$ and $\mathbf{T}_G$ due to the term $\mathbf{FT}_F \mathbf{T}'_G \mathbf{G}'$, it is hard to find the global optimal solution for the problem. Instead, we seek to obtain its local optimal solution by alternatively updating $\mathbf{T}_F$ and $\mathbf{T}_G$ while fixing the other one.

When $\mathbf{T}_G$ is fixed, the objective function w.r.t. $\mathbf{T}_F$ is reduced to

$$J_{\mathbf{T}_F} = \|\tilde{\mathbf{W}} \odot (\tilde{\mathbf{R}} - \mathbf{FT}_F \mathbf{T}'_G \mathbf{G}')\|_F^2 \tag{3.55}$$

$$+ \alpha \operatorname{tr}(\mathbf{T}'_F \mathbf{F}'(\mathbf{D}_{\tilde{M}} - \tilde{\mathbf{M}})\mathbf{FT}_F) + \beta \|\mathbf{T}_F\|_F^2.$$

Then the derivative of $J_{\mathbf{T}_F}$ w.r.t. $\mathbf{T}_F$ is

$$\frac{1}{2}\frac{\partial J_{\mathbf{T}_F}}{\partial \mathbf{T}_F} = \mathbf{F}'(\tilde{\mathbf{W}} \odot \tilde{\mathbf{W}} \odot (\mathbf{FT}_F \mathbf{T}'_G \mathbf{G}'))\mathbf{GT}_G \tag{3.56}$$

$$- \mathbf{F}'(\tilde{\mathbf{W}} \odot \tilde{\mathbf{W}} \odot \tilde{\mathbf{R}})\mathbf{GT}_G + \alpha \mathbf{F}'(\tilde{\mathbf{D}}_{\tilde{M}} - \tilde{\mathbf{M}})\mathbf{FT}_F + \beta \mathbf{T}_F.$$

Therefore, we can update $\mathbf{T}_F$ with

$$\mathbf{T}_F(i, j) = \mathbf{T}_F(i, j) \sqrt{\frac{\mathbf{X}_F(i, j)}{\mathbf{Y}_F(i, j)}}, \tag{3.57}$$

where

$$\mathbf{X}_F = \mathbf{F}'(\tilde{\mathbf{W}} \odot \tilde{\mathbf{W}} \odot \tilde{\mathbf{R}})\mathbf{GT}_G + \alpha \mathbf{F}'\tilde{\mathbf{M}}\mathbf{FT}_F \tag{3.58}$$

$$\mathbf{Y}_F = \mathbf{F}'(\tilde{\mathbf{W}} \odot \tilde{\mathbf{W}} \odot (\mathbf{FT}_F \mathbf{T}'_G \mathbf{G}'))\mathbf{GT}_G \tag{3.59}$$

$$+ \alpha \mathbf{F}'\tilde{\mathbf{D}}_{\tilde{M}}\mathbf{FT}_F + \beta \mathbf{T}_F.$$

Note that the brute force way to update $\mathbf{Y}_F$ requires to calculate a large dense matrix $\tilde{\mathbf{W}} \odot \tilde{\mathbf{W}} \odot (\mathbf{FT}_F \mathbf{T}'_G \mathbf{G}')$ (i.e., $\tilde{\mathbf{W}} \odot \tilde{\mathbf{W}} \odot (\tilde{\mathbf{F}}\tilde{\mathbf{G}}')$). This step will take $O(n_u n_i r)$ which is time-consuming in large-scale systems. Recall that we have set $\tilde{\mathbf{W}}(i, j) = 1$ if $\tilde{\mathbf{R}}(i, j) = 1$ and $\tilde{\mathbf{W}}(i, j) = w$ if $\tilde{\mathbf{R}}(i, j) = 0$, then the above term can be rewritten as $(1 - w^2)\tilde{\mathbf{R}}_e + w^2\tilde{\mathbf{F}}\tilde{\mathbf{G}}'$ where $\tilde{\mathbf{R}}_e = \tilde{\mathbf{R}} \odot$

($\tilde{\mathbf{F}}\tilde{\mathbf{G}}$). In other words, the entries in $\tilde{\mathbf{R}}_e$ are the reconstructed preferences of observed links in the updated preference matrix $\tilde{\mathbf{R}}$, which is very sparse in real applications. Moreover, the term $\tilde{\mathbf{W}} \odot \tilde{\mathbf{W}} \odot \tilde{\mathbf{R}}$ in Eq. (3.58) is equivalent to $\tilde{\mathbf{R}}$ itself. Therefore, the updating rule for $\mathbf{T}_F$ can be simplified as

$$\mathbf{X}_F = (\mathbf{F}'\tilde{\mathbf{R}}\mathbf{G})\mathbf{T}_G + \alpha(\mathbf{F}'\tilde{\mathbf{M}}\mathbf{F})\mathbf{T}_F \tag{3.60}$$

$$\mathbf{Y}_F = (1 - w^2)\mathbf{F}'\tilde{\mathbf{R}}_e\mathbf{G}\mathbf{T}_G + w^2(\mathbf{F}'\mathbf{F})\mathbf{T}_F\mathbf{T}_G'(\mathbf{G}'\mathbf{G})\mathbf{T}_G \tag{3.61}$$
$$+ \alpha(\mathbf{F}'\tilde{\mathbf{D}}_{\tilde{M}}\mathbf{F})\mathbf{T}_F + \beta\mathbf{T}_F.$$

Similarly, $\mathbf{T}_G$ can be updated with

$$\mathbf{T}_G(i,j) = \mathbf{T}_G(i,j)\sqrt{\frac{\mathbf{X}_G(i,j)}{\mathbf{Y}_G(i,j)}}, \tag{3.62}$$

where

$$\mathbf{X}_G = (\mathbf{G}'\tilde{\mathbf{R}}'\mathbf{F})\mathbf{T}_F + \alpha(\mathbf{G}'\tilde{\mathbf{N}}\mathbf{G})\mathbf{T}_G \tag{3.63}$$

$$\mathbf{Y}_G = (1 - w^2)\mathbf{G}'\tilde{\mathbf{R}}_e'\mathbf{F}\mathbf{T}_F + w^2(\mathbf{G}'\mathbf{G})\mathbf{T}_G\mathbf{T}_F'(\mathbf{F}'\mathbf{F})\mathbf{T}_F \tag{3.64}$$
$$+ \alpha(\mathbf{G}'\tilde{\mathbf{D}}_{\tilde{N}}\mathbf{G})\mathbf{T}_G + \beta\mathbf{T}_G.$$

The proposed algorithm is summarized in Algorithm 3.7. It first gets r, the dimension of latent features $\mathbf{F}$ and $\mathbf{G}$ in step 1, and then initializes the transformation matrices $\mathbf{T}_F$ and $\mathbf{T}_G$ randomly in steps 2 and 3. From step 4, the algorithm begins to update $\mathbf{T}_F$ (step 5) and $\mathbf{T}_G$ (step 6) alternatively until convergence.

Algorithm Analysis

We analyze the effectiveness and efficiency of Algorithm 3.7. In terms of the effectiveness of the algorithm, we first show that the fixed point solutions of Eqs. (3.57) and (3.62) satisfy the *KKT* condition.

Theorem 3.13 *The fixed point solutions of Eqs. (3.57) and (3.62) satisfy the KKT condition.*

Proof. As $\mathbf{T}_F$ and $\mathbf{T}_G$ are solved in the same way, we only need to show that the fixed point solution for $\mathbf{T}_F$ in Eq. (3.57) satisfy the *KKT* condition, the other one can be proved in the same procedure.

First, the Lagrangian function for Eq. (3.55) is

$$L_{J_F} = \|\tilde{\mathbf{W}} \odot (\tilde{\mathbf{R}} - \mathbf{F}\mathbf{T}_F\mathbf{T}_G'\mathbf{G}')\|_F^2 + \text{tr}(\mathbf{T}_F'\mathbf{F}'\mathbf{D}_{\tilde{M}}\mathbf{F}\mathbf{T}_F) \tag{3.6}$$
$$- \alpha\text{tr}(\mathbf{T}_F'\mathbf{F}'\tilde{\mathbf{M}}\mathbf{F}\mathbf{T}_F) + \beta\|\mathbf{T}_F\|_F^2 - \text{tr}(\Lambda'\mathbf{T}_F),$$

Algorithm 3.7 ENCORE: The Incremental OCCF Algorithm with Co-Evolving Side Networks

---

Input: (1) The original recommendation input $\Gamma = <\mathbf{M}, \mathbf{N}, \mathbf{R}>$; (2) the perturbations on the system $\Delta\Gamma = <\Delta\mathbf{M}, \Delta\mathbf{N}, \Delta\mathbf{R}>$; (3) the original latent features $\mathbf{F}$ and $\mathbf{G}$; (4) weight w; and (5) regularized parameters α and β

Output: (1) The transformation matrix for user latent features $\mathbf{T}_F$ and (2) the transformation matrix for item latent features $\mathbf{T}_G$

1: $r \leftarrow$ rank of $\mathbf{F}$ and $\mathbf{G}$
2: initialize $\mathbf{T}_F$ as $r \times r$ non-negative random matrix
3: initialize $\mathbf{T}_G$ as $r \times r$ non-negative random matrix
4: **while** not converge **do**
5:    update $\mathbf{T}_F$ as Eq. (3.57)
6:    update $\mathbf{T}_G$ as Eq. (3.62)
7: **end while**
8: return $\mathbf{T}_F, \mathbf{T}_G$

---

where Λ is the Lagrange multiplier. By setting the derivative of L_{J_F} w.r.t. $\mathbf{T}_F$ to 0, we get

$$2(\mathbf{F}'(\tilde{\mathbf{W}} \odot \tilde{\mathbf{W}} \odot (\mathbf{F}\mathbf{T}_F\mathbf{T}'_G\mathbf{G}'))\mathbf{G}\mathbf{T}_G \tag{3.66}$$
$$-\mathbf{F}'(\tilde{\mathbf{W}} \odot \tilde{\mathbf{W}} \odot \tilde{\mathbf{R}})\mathbf{G}\mathbf{T}_G$$
$$+\alpha\mathbf{T}_F\mathbf{F}'\tilde{\mathbf{D}}_{\tilde{M}}\mathbf{F} - \alpha\mathbf{T}_F\mathbf{F}'\tilde{\mathbf{M}}\mathbf{F} + \beta\mathbf{T}_F) = \Lambda.$$

By the KKT slackness condition, we have

$$[-\mathbf{F}'(\tilde{\mathbf{W}} \odot \tilde{\mathbf{W}} \odot \tilde{\mathbf{R}})\mathbf{G}\mathbf{T}_G - \alpha\mathbf{T}_F\mathbf{F}'\tilde{\mathbf{M}}\mathbf{F} \tag{3.67}$$
$$+\mathbf{F}'(\tilde{\mathbf{W}} \odot \tilde{\mathbf{W}} \odot (\mathbf{F}\mathbf{T}_F\mathbf{T}'_G\mathbf{G}'))\mathbf{G}\mathbf{T}_G$$
$$+\alpha\mathbf{T}_F\mathbf{F}'\tilde{\mathbf{D}}_{\tilde{M}}\mathbf{F} + \beta\mathbf{T}_F](i, j)\mathbf{T}_F(i, j) = 0.$$

At the fixed point of Eq. (3.57), we have $\mathbf{X}_F(i, j) = \mathbf{Y}_F(i, j)$, which implies that

$$[\mathbf{F}'(\tilde{\mathbf{W}} \odot \tilde{\mathbf{W}} \odot \tilde{\mathbf{R}})\mathbf{G}\mathbf{T}_G + \alpha\mathbf{T}_F\mathbf{F}'\tilde{\mathbf{M}}\mathbf{F}](i, j) \tag{3.68}$$
$$= [\mathbf{F}'(\tilde{\mathbf{W}} \odot \tilde{\mathbf{W}} \odot (\mathbf{F}\mathbf{T}_F\mathbf{T}'_G\mathbf{G}'))\mathbf{G}\mathbf{T}_G$$
$$+ \alpha\mathbf{T}_F\mathbf{F}'\tilde{\mathbf{D}}_{\tilde{M}}\mathbf{F} + \beta\mathbf{T}_F](i, j).$$

Clearly, the above equation satisfies the KKT slackness condition in Eq. (3.67). Therefore, the fixed point solution of Eq. (3.57) satisfies the KKT condition. $\square$

Theorem 3.13 states that the updating rules in Eq. (3.57) lead to a local optimal solution to Eq. (3.55) at convergence. Also, it can be proved that by following the updating rule in Eq. (3.57), the objective function in Eq. (3.55) decreases monotonically.

Combining Theorem 3.13 with the monotonic decreasing property, we can conclude that Algorithm 3.7 converges to the local optimal solution $\mathbf{T}_F$ for the objective function in Eq. (3.55). Similarly, we have the local optimal solution $\mathbf{T}_G$. The two matrices $\mathbf{T}_F$ and $\mathbf{T}_G$ form the local optimal solution for Eq. (3.54).

For efficiency of the algorithm, we analyze the time complexity and space complexity of Algorithm 3.7 in Lemmas 3.14 and 3.15, respectively.

Lemma 3.14   *The time complexity of proposed algorithm is* $O((\tilde{m}_u + \tilde{m}_i)r + ((n_u + n_i + r)r^2 + \tilde{m}_r r)t)$.

Proof. In Algorithm 3.7, as term $\tilde{\mathbf{M}}$, $\tilde{\mathbf{N}}$, $\tilde{\mathbf{R}}$, $\mathbf{F}$, and $\mathbf{G}$ remain the same during the iterations, we can pre-compute related constant terms to avoid redundant computations. The complexities of computing constant terms in Eqs. (3.60)–(3.64) are $O(n_i r^2 + \tilde{m}_r r)$ for $\mathbf{F}'\tilde{\mathbf{R}}\mathbf{G}$; $O(n_u r^2 + \tilde{m}_u r)$ for $\mathbf{F}'\tilde{\mathbf{M}}\mathbf{F}$; $O(n_u r^2)$ for $\mathbf{F}'\mathbf{D}_{\tilde{M}}\mathbf{F}$; $O(n_u r^2)$ for $\mathbf{F}'\mathbf{F}$; $O(n_i r^2)$ for $\mathbf{G}'\mathbf{G}$; $O(n_i r^2 + \tilde{m}_i r)$ for $\mathbf{G}'\tilde{\mathbf{N}}\mathbf{G}$; and $O(n_i r^2)$ for $\mathbf{G}'\mathbf{D}_{\tilde{N}}\mathbf{G}$. Thus, the complexity for pre-computing is $O((n_u + n_i)r^2 + (\tilde{m}_u + \tilde{m}_i + \tilde{m}_r)r)$. In each iteration, it takes $O(n_u r^2)$ and $O(n_i r^2)$ to compute $\mathbf{FT}_F$ (i.e., $\tilde{\mathbf{F}}$) and $\mathbf{GT}_G$ (i.e., $\tilde{\mathbf{G}}$), respectively. The complexity of computing $\mathbf{F}'\tilde{\mathbf{R}}_e\mathbf{GT}_G$ is $O(n_i r^2 + \tilde{m}_r r)$, the rest of the computations for updating $\mathbf{T}_F$ and $\mathbf{T}_G$ are both $O(r^3)$. Therefore, the overall complexity for Algorithm 3.7 is $O((\tilde{m}_u + \tilde{m}_i)r + ((n_u + n_i + r)r^2 + \tilde{m}_r r)t)$, where t is the number of iterations in the algorithm.   □

Compared with the complexity of static OCCF algorithm in the previous section $(O(((m_u + m_i + m_r)r + (n_u + n_i)r^2)t))$, the proposed ENCORE is more efficient, with an $O((\tilde{m}_u + \tilde{m}_i)r)$ reduction in the time complexity in *each* iteration.

Lemma 3.15   *The space complexity of proposed algorithm is* $O((n_u + n_i + r)r + \tilde{m}_u + \tilde{m}_i + \tilde{m}_r)$.

Proof. The algorithm requires a space of $O(n_u r + n_i r)$ to store $\mathbf{F}$ and $\mathbf{G}$, $O(r^2)$ to store transformation matrices $\mathbf{T}_F$ and $\mathbf{T}_G$, and $O(\tilde{m}_u + \tilde{m}_i + \tilde{m}_r)$ to store the updated rating matrix and side networks. The space needed to compute and store the constant terms are $O(n_i r + r^2)$ for $\mathbf{F}'\tilde{\mathbf{R}}\mathbf{G}$; $O(n_u r + r^2)$ for $\mathbf{F}'\tilde{\mathbf{M}}\mathbf{F}$ and $\mathbf{F}'\mathbf{D}_{\tilde{M}}\mathbf{F}$; $O(r^2)$ for $\mathbf{F}'\mathbf{F}$ and $\mathbf{G}'\mathbf{G}$; $O(n_i r + r^2)$ for $\mathbf{G}'\tilde{\mathbf{N}}\mathbf{G}$ and $\mathbf{G}'\mathbf{D}_{\tilde{N}}\mathbf{G}$. Therefore, the space costs for computing constant terms is $O((n_u + n_i)r + r^2)$. In each iteration, it takes a space of $O((n_u + n_i)r)$ to compute $\tilde{\mathbf{F}}$ and $\tilde{\mathbf{G}}$, $O(\tilde{m}_r)$ to compute $\tilde{\mathbf{R}}_e$, $O(n_i r + r^2)$ to compute $\mathbf{F}'\tilde{\mathbf{R}}_e\mathbf{GT}_G$, $O(r^2)$ for the rest of the matrix multiplications to update $\mathbf{T}_F$ and $\mathbf{T}_G$. Putting all together, the overall space complexity for Algorithm 3.7 $O((n_u + n_i + r)r + \tilde{m}_u + \tilde{m}_i + \tilde{m}_r)$.   □

Variations

In this section, we discuss some of the variants of ENCORE. First, when the weighting matrix $\mathbf{W}$ is an all-one matrix, ENCORE becomes a dynamic clustering algorithm, in which $\mathbf{T}$

and $\mathbf{T}_G$ can be viewed as the cluster membership transition matrix. Second, when one or both sides networks are missing, the corresponding regularization term would be removed from the objective function. In particular, when both sides of the networks were removed, ENCORE is reduced to a dynamic algorithm for the classic one-class collaborative filtering problem.

3.3.3   EXPERIMENTAL EVALUATIONS

In this section, we evaluate the proposed ENCORE algorithm on two real datasets. The experiments are designed to answer the following two questions.

- *Effectiveness*. How effective is ENCORE for OCCF problem with co-evolving side networks?

- *Efficiency*. How fast is ENCORE compared with batch-mode static counterpart?

Experimental Setup
We first introduce the datasets used, comparing methods, evaluation metrics, and experimental settings before presenting the details of the experiments.

Datasets Description. We use two real datasets Ciao [118] and Epinions [119] to evaluate the proposed ENCORE method. Ciao and Epinions are two popular online product review websites in which users are allowed to build connections and share experiences on the products with each other. To fit the one-class collaborative filtering problem, all missing links and ratings that are no greater than 3 are viewed as negative examples (i.e., labeled as 0), while ratings that are greater or equal to 4 are marked as positive examples (i.e., labeled as 1). The user side network contains the trust relations between users, while the item side network describes the similarity between items based on their reviews.[6] Both datasets have been preprocessed and used in [139] and are publicly available. The statistics of the datasets are summarized in Table 3.11.

In our evaluation, the datasets are partitioned into three groups. The first group is the original training system which contains 50% of the ratings and the corresponding side network links; the second group is the incremental system, which adds 1% links to the rating matrix and the side network connections at each time stamp; the last group is the testing system, which contains the rest of the data.

Comparing Methods. We compare ENCORE with the following baseline methods to demonstrate its effectiveness.

- **ReRun.** *ReRun* is the batch-mode static counterpart for ENCORE. At each time stamp, it takes the current system snapshot as input networks and solves the optimization problem in Eq. (3.52) from scratch.[7] As *ReRun* does not impose any transformation constraints on the

---

[6]Similarity between items is calculated by the cosine similarity between TF-IDF word vectors constructed from item reviews.

[7]Equation (3.52) is derived from wiZAN-Dual in [139]. Therefore, ReRun is equivalent to wiZan-Dual here.

Table 3.11: Statistics of datasets

Dataset	Ciao	Epinions
Number of users	6,102	33,725
Number of items	12,082	43,542
Number of user links	75,861	328,455
Number of items links	283,284	249,397
Number of preferences	117,731	500,478
Mean degree of users	24.86	19.48
Cluster coefficient of users	0.13	0.10
Mean degree of items	23.45	11.46
Cluster coefficient of items	0.72	0.35

latent features in two consecutive time stamps, it can be used to validate the effectiveness of the transformation model in ENCORE.

- **M+R**. *M+R* is a variant of ENCORE, which only contains user side network and preference matrix in the system.

- **N+R**. *N+R* is another variant of ENCORE, which only contains item side network and preference matrix.

- **R-MF**. *R-MF* is a simple method for OCCF proposed in [95], which only utilizes the preference matrix in the system for the recommendation.

- **CLiMF**. *CLiMF* is also a matrix factorization based method proposed in [111] that is designed to improve the performance of top-k recommendations on binary relevance data.

- **R-SGD**. *R-SGD* shares the same objective function with *R-MF*. Instead of calculating the latent features at each time stamp, *R-SGD* modifies related latent features with newly emerged ratings by stochastic gradient descent method.

- **eNMF**. *eNMF* is an incremental matrix factorization algorithm proposed in [128]. Here we apply this algorithm to perform incremental update on the factorization results of the preference matrix.

Evaluation Metrics. In our experiments, we assess the effectiveness of ENCORE with *MA* and *R-MPR* as evaluation metrics.

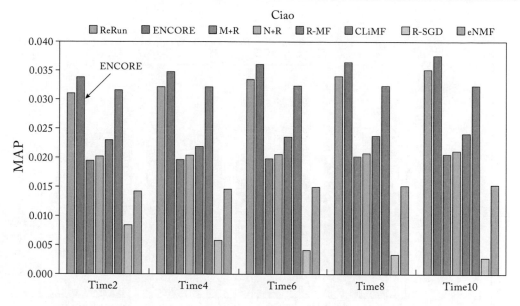

Figure 3.26: MAP comparison on ciao. Higher is better. (Best viewed in color.)

- **MAP**. *MAP* (Mean Average Precision) is originally used to evaluate ranked documents over a set of queries. Here it computes the mean average precision over all users in the test set [95]. The larger the *MAP* is, the better the performance is.

- **R-MPR**. *R-MPR* (Reverse Mean Percentage Ranking) is a variation of *MPR*, which is originally used to evaluate users' satisfaction of items by a ranked list. A randomly generated item list can achieve an *MPR* of 50% [75]. The smaller the *MPR* is, the better the performance is. Here we set *R-MPR* to be 0.5-*MPR*, thus a larger *R-MPR* indicates better performance.

Machine. The experiments are performed on a machine with 2 Intel Xeon 3.5 GHz processors and 256 GB of RAM. The algorithms are implemented with MATLAB using a single thread.

Effectiveness Results

We compare the proposed algorithm with other methods on both Ciao and Epinions datasets. The results are shown in Figures 3.26–3.29. We make the following observations from these two figures.

- In both datasets, ENCORE achieves close performance with *ReRun*. Such results demonstrate that ENCORE can effectively accommodate newly emerged links in the dynamic system for the recommendation. It should be noted that here our goal is not to develop a better recommendation algorithm that outperforms *ReRun*. Instead, our goal is to ensure that the

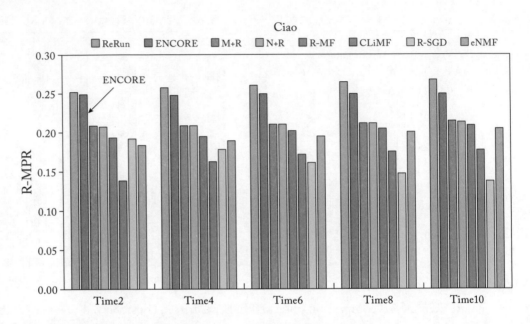

Figure 3.27: R-MPR comparison on ciao. Higher is better. (Best viewed in color.)

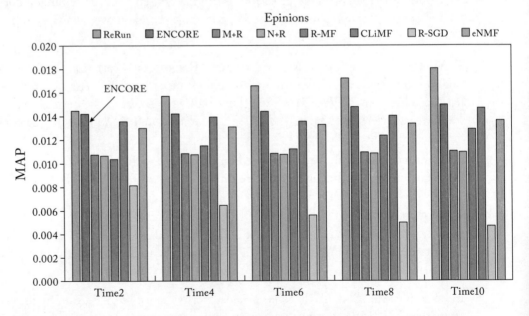

Figure 3.28: MAP comparison on epinions. Higher is better. (Best viewed in color.)

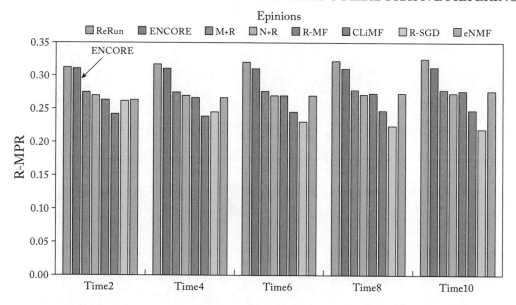

Figure 3.29: R-MPR comparison on epinions. Higher is better. (Best viewed in color.)

incrementally updated model with linear transformation at each time stamp can closely approximate the performance of *ReRun*.

- Side networks between users and items are both important for improving the quality of recommendation results. As we can see from Figures 3.26–3.29, a user network or item network alone with the preference matrix cannot boost the recommendation quality compared with the methods that use preference matrix only. However, when both networks are added to the model, the performance can be improved significantly.

- For the two incremental update algorithms (ENCORE and eNMF), our proposed algorithm ENCORE is consistently better than eNMF due to its effectiveness in exploiting the information from side networks.

- The CLiMF method is designed to maximize the Mean Reciprocal Rank (MRR) for top-k recommendations. Therefore, its MAP score is comparable to ENCORE on both datasets as in Figures 3.26 and 3.28. However, such a scheme would ignore the recall on the recommended items, which results in lower R-MPR scores as indicated in Figures 3.27 and 3.29.

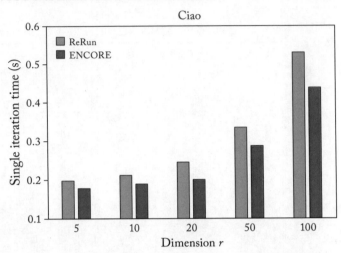

Figure 3.30: The running time of *ReRun* vs. ENCORE for a single iteration on the ciao dataset.

Efficiency Results

We evaluate the efficiency of ENCORE on both Ciao and Epinions. As the results are similar, we only report the one on Ciao for brevity. In the experiment, we have varied the dimension of latent features r with the same input data and evaluated the running time ENCORE under different settings. It can be seen from Figure 3.30 that the average running time of ENCORE for a single iteration is shorter than the *ReRun* method. Specifically, as the latent feature dimension r increases, the speed-up of ENCORE compared to *ReRun* becomes larger accordingly. This observation is consistent with our time complexity analysis in the previous section that the proposed ENCORE algorithm has an $O(r)$ factor speed-up over its static counterpart (*ReRun*). Moreover, as shown in Figure 3.31, the average running time of ENCORE for one time stamp is much shorter than *ReRun* (with around 75% improvement). This is mainly due to the fact that ENCORE has much fewer variables to optimize at each time stamp compared to *ReRun* (i.e., $2r \times r$ vs. $(m + n) \times r$), which makes it converge faster in a small number of iterations.

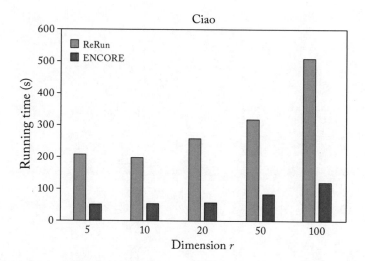

Figure 3.31: The running time of *ReRun* vs. ENCORE for one time stamp (many iterations until convergence) on the ciao dataset.

CHAPTER 4

Network Connectivity Optimization

Network connectivity optimization plays a critical role in many applications. In the immunization studies [28], it is critical to select a group of nodes/edges to effectively contain the propagation process [17, 19]. In the biomedical domain, antibiotic drugs are developed to kill the bacteria by disrupting their molecular network to the max extend [58]. While in the critical infrastructure networks, facilities that may cause large-scale failures are retrieved and protected proactively to ensure the full-functioning of the entire system [22]. It is worth noting that finding a group of nodes/edges that have high impact on the connectivity of the network is similar to the influence maximization problem [26, 53, 86] and its variations (e.g., viral marketing [25, 64], outbreak detection [67], etc.). The main difference between the two problems is that the influence maximization problem is highly dependent on the underlying diffusion model (e.g., independent cascade, linear threshold [53]), while the connectivity optimization problem is directly based on the backbone (i.e., the underlying topology) of the network. In addition to the global connectivity measures, the local connectivity of the network has also been applied to some critical tasks like graph clustering [140, 141], link prediction [6], etc.

To effectively manipulate the connectivity of the network, earlier work, either explicitly or implicitly, assumes that nodes/edges with higher centrality scores would have a greater impact on network connectivity. This assumption has led to many research efforts on finding good node/edge centrality measures (or node/edge importance measures in general). Some widely used centrality measures include shortest path based centrality [38], PageRank [94], HITS [57], coreness score [85], local Fiedler vector centrality [23], and random walks based centrality [89]. Different from those node centrality-oriented methods, some recent work aims to take one step further by *collectively* finding a subset of nodes/edges with the highest impact on the network connectivity measure. For example, [18, 19, 123, 124] proposed both node-level and edge-level manipulation strategies to optimize the leading eigenvalue of the network, which is the key network connectivity measure for a variety of cascading models. In [14], Chan et al. further generalized these strategies to manipulate the network robustness scores [51]. In [16], Chen et al. showed that the connectivity optimization problem is generally NP-hard and proposed an efficient algorithm for both node- and edge-level optimization tasks.

In this section, we introduce the optimization strategies for SUBLINE connectivity measures and their extensions to multi-layered networks.

4.1 SUBLINE CONNECTIVITY OPTIMIZATION

As we have discussed in Chapter 2, the connectivity of the network may take different forms depending on the types of the applications. Despite the various forms, the connectivity minimization problem has always been a fundamental task in most applications, in which a less connected network or subnetwork is more preferred [60]. The primary goal for connectivity minimization is to find a set of nodes/edges whose removal may lead to the destruction of the underlying network. For example, in the critical infrastructure construction scenario, the high-impact facilities and links identified by the connectivity minimization algorithms can be viewed as the backbone of the network, which is essential to ensure the full functioning of the entire system. While in the immunization scenario, disease control centers need to vaccine high impact entities and cut down highly contagious connections to prevent the prevalence of the disease.

The main computation obstacle for the connectivity minimization problems lies in its combinatorial nature. Specifically, for the global connectivity minimization problem, suppose the number of nodes and edges in the network is n and m, respectively, then the number of all possible node sets of size k would be $\binom{n}{k}$ and the number of all possible edge sets is $\binom{m}{k}$. Such exponential complexity would make exhaust search intractable even in mid-sized networks.

To reduce the exponential time complexity, existing algorithms predominantly rely on the greedy scheme. Taking the node deletion-based connectivity minimization problem as an example, the greedy scheme would iteratively collect the node that has the largest impact on the pre-defined connectivity in the network until the budget is used up. By virtue of the diminishing returns property on a wide range of the connectivity minimization problems [15], the greedy scheme can secure a near-optimal approximated solution with an approximation ratio of $1 - 1/e$ [88]. A key step in the greedy scheme is to calculate the impact of each candidate node/edge on the given connectivity measure, which often involves eigendecomposition operations with *polynomial* complexity w.r.t. the size of the network. Obviously, a polynomial algorithm still cannot handle large-scale networks efficiently. To further accelerate the algorithm, matrix perturbation-based methods are frequently used to approximate the impact of a node/edge [18]. Such approximation algorithms have been proved to scale linearly w.r.t. the network size, while exhibiting empirical superiority over other alternative methods.

Although the above-mentioned methods are empirically effective for some specific connectivity minimization problems, two main challenges for the general connectivity minimization problem still remain largely open. On the theoretical side, the *hardness* of the general connectivity minimization problem has never been systematically justified except for a few special instances (e.g., epidemic threshold [19] and triangle capacity [74]). Furthermore, although the greedy scheme can guarantee a $1 - 1/e$ approximation ratio for the connectivity minimization problem, it still remains unknown if a better approximation ratio can be achieved within polynomial time. On the algorithmic side, exact greedy algorithms often bear polynomial time complexity which is not scalable in large-scale networks. Although matrix perturbation-based approximation methods can simplify the complexity down to the linear scale, their optimization quality highly dependent on the spectrum of the underlying network (the optimization quality would deteriorate quickly in networks with small eigen-gaps [17, 62]).

In this work, we address the theoretical and algorithmic challenges of the connectivity minimization problem. The main contributions of this work can be summarized as follows.

- *Revealing the Fundamental Limits.* We prove that for the connectivity minimization problem on a wide range of connectivity measures (1) is NP-complete and (2) $(1 - 1/e)$ is the best approximation ratio for any polynomial algorithms, unless $NP \subseteq DTIME(n^{O(\log \log n)})$.[1]

- *Developing New Algorithms.* We propose an effective algorithm (CONTAIN) for network connectivity optimization. The centerpieces of the proposed method include (a) an effective impact score approximation method and (b) an efficient eigen-pair update method. The proposed CONTAIN algorithm bears three distinct advantages over the existing methods, including (1) *effectiveness*, being able to handle small eigen-gap networks, consistently outperforming the state-of-the-art methods over a diverse set of real networks; (2) *scalability*, with a linear complexity w.r.t. the network size; and (3) *generality*, applicable to a variety of different network connectivity measures (e.g., leading eigenvalue, triangle capacity, and natural connectivity) as well as network operations (node vs. edge deletion). In addition, we also propose a variation of CONTAIN (CONTAIN+) which can further simply the computational complexity by deriving a closed-form approximation on node/edge impact scores.

4.1.1   PROBLEM DEFINITION

In this section, we formally introduce the network connectivity optimization problem and review the general strategy of greedy algorithms.

Table 4.1 gives the main symbols used in this work. Following the convention, we use bold uppercase for matrices (e.g., $\mathbf{A}$), bold lowercase for vectors (e.g., $\mathbf{a}$) and calligraphic for sets (e.g., $\mathcal{A}$). We use $\tilde{\ }$ to denote the notations after node/edge deletion, and Δ to denote the perturbations (e.g., $\Delta \mathbf{A} = \tilde{\mathbf{A}} - \mathbf{A}$). $C(G)$[2] represents the network connectivity measure to be optimized in G; o indicates an element (a node/edge) in network G; $I(o)$ denotes the impact score of element o on $C(G)$; $\mathbf{\Lambda}$ and $\mathbf{U}$ denote the eigenvalue matrix and eigenvector matrix for the adjacency matrix $\mathbf{A}$ of the network.

Recall that in Chapter 2 we define the SUBLINE connectivity as

$$C(G) = \sum_{\pi \in G} f(\pi), \tag{4.1}$$

where π is a subgraph of G, f is a non-negative function that maps any subgraph in G to a non-negative real number (i.e., $f : \pi \to \mathbb{R}^+$). Specifically, we have $f(\phi) = 0$ for empty set ϕ; when $f(\pi) > 0$, we call subgraph π as a *valid subgraph*. In other words, the network connectivity

---

[1]$DTIME(t(n))$: the collection of languages that are decidable by $O(t(n))$ time-deterministic Turing machine [113].
[2]$C(G)$ is the abbreviation for $C(G, f)$, which is related to both the network structure and connectivity function f.

Table 4.1: Main symbols for CONTAIN

Symbol	Definition and Description
$G(V, E)$	an undirected network
$\mathbf{A}, \mathbf{B}$	the adjacency matrices (bold uppercase)
$\mathbf{a}, \mathbf{b}$	column vectors (bold lowercase)
$\mathcal{A}, \mathcal{B}$	sets (calligraphic)
$\mathbf{A}(i, j)$	the element at the ith row and the jth column in $\mathbf{A}$
$\mathbf{a}(i)$	the ith element of vector $\mathbf{a}$
$\mathbf{A}'$	transpose of matrix $\mathbf{A}$
$\Delta \mathbf{A}$	perturbation of $\mathbf{A}$
$\tilde{\mathbf{A}}$	the adjacency matrix after node/edge deletion on $\mathbf{A}$
m, n	number of edges and nodes in network G
$C(G)$	connectivity measure of network G under mapping function f
$C_{\mathcal{T}}(G)$	the local connectivity of subgraph $\mathcal{T}$ on network G
$F(\Lambda^{(r)})$	associated eigen-function for $C(G)$
o	a network element in G (a node/edge)
$I(o)$	connectivity impact score of o on $C(G)$
$I_{\mathcal{T}}(o)$	local connectivity impact score of o on $C_{\mathcal{T}}(G)$
$\lambda, \mathbf{u}$	the leading eigenvalue and eigenvector of $\mathbf{A}$ (in magnitude)
$\Lambda, \mathbf{U}$	the eigenvalue and eigenvector matrix of $\mathbf{A}$
$\Lambda^{(r)}, \mathbf{U}^{(r)}$	the top-r eigen-pairs of $\mathbf{A}$ (in magnitude)
k	the budget

$C(G)$ can be viewed as a weighted aggregation of the connectivities of all valid subgraphs in the network. When the valid subgraphs are defined on a subset of nodes $\mathcal{T}$, the connectivity measure defined in Eq. (4.1) can also be extended to measure the local connectivity of a subset of node $\mathcal{T}$, where we define $f(\pi) > 0$ iff π is incident to the node set $\mathcal{T}$. Moreover, by choosing a appropriate $f()$ function (refer to Chapter 2 for details), Eq. (4.1) includes several prevalent network connectivity measures, e.g., path capacity, triangle capacity, and natural connectivity. In terms of computation, it is often much more efficient to either approximate or compute these connectivity measures by the associated eigen-function $F(\Lambda^{(r)})$, where $\Lambda^{(r)}$ represents the top-eigenvalues of $\mathbf{A}$.

Network Connectivity Minimization

With the network connectivity measure in Eq. (4.1), we formally define the network connectivity optimization problem as follows.

Problem 4.1   Network Connectivity Minimization (NETCOM)

Given: (1) A network G; (2) a connectivity mapping function $f : \pi \to \mathbb{R}^+$ which defines $C(G)$; (3) a type of network operation (node deletion vs. edge deletion); and (4) an integer budget k with $1 < k < \min\{|\mathcal{S}_\pi|, K\}$, where $\mathcal{S}_\pi = \{\pi \,|\, f(\pi) > 0\}$ denotes the set of valid subgraphs and K denotes the number of valid network elements.

Output: A set of network elements $\mathcal{X}$ of size k, whose removal from G would minimize connectivity $C(G)$.

It is worth noting that depending on the definition of $C(G)$, the valid subgraphs in $\mathcal{S}_\pi$ may have various structures. In the triangle minimization scenario, $\mathcal{S}_\pi$ contains all the triangles in the network. When the valid subgraph shares the same form as the operation type (i.e., a valid subgraph is a single node in the node-level operation scenario, or a valid subgraph is an edge in the edge-level operation scenario), we call this kind of valid subgraph as *singleton*. In Problem 4.1, we also require that the budget $1 < k < \min\{|\mathcal{S}_\pi|, K\}$. This is a fairly generic constraint that can be easily met. For example, for the node deletion operation, the set of valid network elements is simply the entire node set of the input network (i.e., $K = n$); for a connected network with its connectivity measure $C(G)$ defined as the path capacity, we have that $|\mathcal{S}_\pi| > n$. Therefore, the above constraint simply means that we cannot delete all the nodes from the input network, which would make the problem trivial. On the other end of the spectrum, we require that the budget $k > 1$. Otherwise (with $k = 1$), the problem can be easily solved in polynomial time (e.g., by choosing the valid network element with the largest impact score). Problem 4.1 provides a general definition of the network connectivity optimization problem, which can be in turn instantiated into different instances, depending on (1) the specific choice of the connectivity measure $C(G)$ (or equivalently the choice of the $f()$ function) and (2) the type of network operation (node deletion vs. edge deletion). For example, in the robustness analysis of the power grid, we might choose the natural connectivity as $C(G)$ to evaluate the robustness of the system, and we are interested in identifying k most critical power transmission lines whose failure would cause a cascading failure of the entire grid. To abstract it as a network connectivity optimization problem, we have the input network set as the topological structure of the power grid; the connectivity to optimize as the natural connectivity; the operation type as edge deletion; and the valid network elements as all the edges (i.e., $K = m$ in this case).

The NETCOM problem can be easily extended to local connectivity measures. Specifically, the local connectivity minimization problem can be defined as follows.

Problem 4.2   Local Connectivity Minimization

Given: (1) A network G; (2) a subset of target nodes $\mathcal{T}$; (3) a connectivity mapping function $f :$ $\pi \to \mathbb{R}^+$ which defines the local connectivity measure $C_{\mathcal{T}}(G)$; (4) a type of network operation (node deletion vs. edge deletion); and (5) an integer budget k with $1 < k < \min\{|\mathcal{S}_\pi|, K\}$, where $\mathcal{S}_\pi = \{\pi \mid f(\pi) > 0\}$ denotes the set of valid subgraphs and K denotes the number of valid network elements.

Output: A set of network elements $\mathcal{X}$ of size k with $\mathcal{X} \cap \mathcal{T} = \Phi$, whose removal from G would minimize connectivity $C_{\mathcal{T}}(G)$.

Note that in Problem 4.2, the restriction $\mathcal{X} \cap \mathcal{T} = \Phi$ on $\mathcal{X}$ is used to avoid the trivial solution under node deletion operations, in which target nodes $\mathcal{T}$ are removed for local connectivity minimization.

Greedy Strategy for NETCOM

Due to the combinatorial nature of Problem 4.1, it is computationally infeasible to solve it in a brute-force manner. Thanks to the diminishing returns property of NETCOM, the greedy strategy has become a prevalent choice for solving Problem 4.1 with a guaranteed $(1 - 1/e)$ approximation ratio. For the ease of following discussions, we present the outline of such greedy strategy in Algorithm 4.8. In Algorithm 4.8, the solution set $\mathcal{X}$ is initialized with an empty set. At each iteration (steps 2–8), the element (a node or an edge) with the highest impact score is added to the solution set $\mathcal{X}$ until the budget is reached. The returned solution set $\mathcal{X}$ in step 9 guarantees a $(1 - 1/e)$ approximation ratio. For more details and proofs, please refer to [15].

Algorithm 4.8 A Generic Greedy Strategy for NETCOM [15]

Input: (1) A network G; (2) a connectivity mapping function $f : \pi \to \mathbb{R}^+$ which defines $C(G)$; (3) a type of network operation; and (4) a positive integer k

Output: A set of network elements $\mathcal{X}$ of size k

1: initialize $\mathcal{X}$ to be empty
2: **for** $i = 1$ to k **do**
3:     **for** each valid network element o in G **do**
4:         calculate $I(o) \leftarrow C(G) - C(G \setminus \{o\})$
5:     **end for**
6:     add the element $\tilde{o} = \text{argmax}_o I(o)$ to $\mathcal{X}$
7:     remove the element $\{\tilde{o}\}$ from network G
8: **end for**
9: return $\mathcal{X}$

4.1.2   FUNDAMENTAL LIMITS

In this section, we start with detailing the theoretic challenges of the network connectivity optimization (NETCOM) problem and then reveal two fundamental limits, including its hardness and its approximability.

Theoretic Challenges of NETCOM

The first theoretic challenge of NETCOM lies in its hardness. Since the NETCOM problem has various instances, intuitively, the hardness of those instances might vary dramatically from one to another. For example, if the elements in the valid subgraph set S_π are all singletons w.r.t. the corresponding operation type (i.e., S_π is the node set of the input network for the node-level optimization problem, or S_π is the edge set for the edge-level optimization problem), we can simply choose the top-k nodes/edges with the highest $f(\pi)$ scores, which immediately gives the optimal solution. However, if NETCOM is instantiated as an edge minimization problem under node deletion operations (i.e., the valid subgraph S_π consists of all the edges, the valid network element set is the entire node set), the problem would become the (weighted) max-k vertex cover problem, which is known to be NP-hard. Such observations naturally give rise to the following question, *what is the key intrinsic property of valid subgraph set S_π in conjunction with the network operation type that determines whether or not the corresponding NETCOM instance is polynomially solvable?* To date, the hardness of the general NETCOM problem has largely remained unvalidated, except for a few special instances. The second theoretic challenge of NETCOM lies in its approximability. The greedy algorithm outlined in Section 4.1.1 has a provable $(1 - 1/e)$ approximation ratio [15]. However, we still do not know if such an approximation ratio is *optimal*. In other words, it remains unknown if there exists any polynomial algorithm with an approximation ratio better than $(1 - 1/e)$ for NETCOM.

Fundamental Limit #1: NP-Completeness

We reveal the hardness result of the NETCOM problem in Theorem 4.3. It states that the NETCOM problem defined in Problem 4.1 is in general NP-complete, unless the valid subgraphs in set S_π are mutually independent to each other.[3]

Theorem 4.3   NP-Completeness of NETCOM. *The* NETCOM *problem with non-independent valid subgraphs in Problem 4.1 is NP-complete.*

Proof. As the NETCOM problem admits two possible network operations, including node deletions and edge deletions, we present our proof for each scenario in the following two lemmas. Lemma 4.4 together with Lemma 4.7 would prove that NETCOM problem is NP-complete. □

---

[3]Two valid subgraphs are independent to each other if they do not have any common valid network element.

Lemma 4.4   *The k-node connectivity minimization problem is NP-complete.*

Proof. By Eq. (4.1), the connectivity of network G is defined as $C(G) = \sum_{\pi \in G} f(\pi)$. We set function f as

$$f(\pi) = \begin{cases} w_\pi > 0 & \text{if } \pi \text{ is a valid subgraph} \\ 0 & \text{otherwise.} \end{cases} \tag{4.2}$$

Hence, we formulate the k-node minimization problem as follows.

Problem 4.5 k-**Node Minimization Problem:** $NodeMin(G, k)$

Given: (1) A network $G =< V, E >$; (2) the connectivity function f as defined in Eq. (4.2); and (3) the budget k.

Output: A set with k nodes $V' \subseteq V$, such that $C(G \setminus V')$ (i.e., the connectivity in $G \setminus V'$) is minimized.

Here we first prove that $NodeMin(G, k)$ is NP-hard by constructing a polynomial reduction from a well-known NP-hard problem, the max k-coverage problem (i.e., $MaxCover(n, m, k)$) [56]. The $MaxCover(n, m, k)$ problem is defined as follows.

Problem 4.6   **Max k-Coverage Problem:** $MaxCover(n, m, k)$

Given: (1) The universal set of n elements $\mathcal{U} = \{e_1, e_2, \ldots, e_n\}$; (2) a collection $\mathcal{S} = \{\mathcal{B}_1, \ldots, \mathcal{B}_m\}$ of m distinct subsets of $\mathcal{U}$, which are not mutually exclusive; (3) the non-negative weights $W = \{w_i, \ldots, w_n\}$ associated to the corresponding elements in $\mathcal{U}$; and (4) a positive integer k.

Output: A set $\mathcal{S}' \subseteq \mathcal{S}$ with $|\mathcal{S}'| \leq k$, s.t. $\sum_{e_i \in \mathcal{U}'} w_i$ is maximized, where $\mathcal{U}'$ is the set of elements covered by the sets in $\mathcal{S}'$.

We aim to prove that $MaxCover(n, m, k)$ is polynomially reducible to the k-node minimization problem $NodeMin(G, k)$ (i.e., $MaxCover(n, m, k) \leq_p NodeMin(G, k)$). Without loss of generality, we assume that $1 < k < m$. The rationality behind this assumption is that when $k = 1$ $MaxCover(n, m, 1)$ can be trivially solved by picking the set in $\mathcal{S}$ that contains the elements with the largest weight sum (i.e., S'=$\{\arg\max_{B \in \mathcal{S}} \sum_{e_i \in B} w_i\}$), while for $k \geq m$, we may just take all the subsets in $\mathcal{S}$ into $\mathcal{S}'$ to guarantee a maximum coverage.

Given an instance of $MaxCover(n, m, k)$ with $1 < k < m$, we can construct a network with m nodes, each corresponding to one subset in $\mathcal{S}$. For each element e_i in $MaxCover(n, m, k)$ we construct a valid subgraph G_i as follows. First, we scan set $\mathcal{S}$ and obtain all the set $\{\mathcal{B}_1^i, \ldots, \mathcal{B}_l^i\} \subseteq \mathcal{S}$ that contain element e_i. Then we map $\{\mathcal{B}_1^i, \ldots, \mathcal{B}_l^i\}$ into the nodes in G an

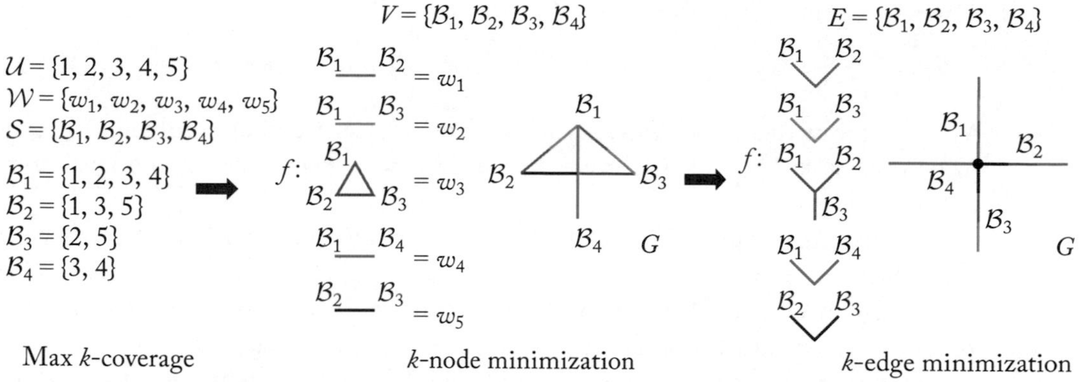

Figure 4.1: An illustration of polynomial reduction from max k-coverage problem.

get the corresponding l nodes. By connecting those l nodes with edges, we get a subgraph G_i in G with connectivity score $f(G_i) = w_i$. In this way, removing any nodes from G_i would break the subgraph completeness. Repeating the above process for all the elements in $\mathcal{U}$, the final graph we get is $G = G_1 \cup \ldots \cup G_n$, and the connectivity function is defined as $f(G_i) = w_i$. Since the sets in $\mathcal{S}$ are distinct and not mutually exclusive, the resulting valid subgraphs are guaranteed to be non-independent. Therefore, the solution of $MaxCover(n, m, k)$ would be equivalent to the solution of $NodeMin(G, k)$, which proves that $NodeMin(G, k)$ is NP-hard.

On the other hand, given a candidate solution set of k node, we can verify its optimality by comparing to the $\binom{n}{k}$ all possible solution sets of size k. As calculating the impact of each set take polynomial time w.r.t. the network size and $\binom{n}{k}$ is also polynomial to the size of the network, the overall complexity to verify the optimality of the candidate solution is also polynomial to the size of the network, which indicates that $NodeMin(G, k)$ is in NP.

Therefore, the $NodeMin(G, k)$ problem is NP-complete. $\qquad\square$

Figure 4.1 gives an illustration of the reduction from an instance of max k-coverage problem $MaxCover(n, m, k)$ to k-node minimization problem $NodeMin(G, k)$, in which different valid subgraphs are marked with different colors. Edges with multiple colors indicate that they are involved in multiple valid subgraphs.

Lemma 4.7 *The k-edge connectivity minimization is NP-complete.*

Proof. We still use the connectivity measure defined Eq. (4.2) to complete the proof. The corresponding k-edge minimization problem can be defined as follows.

Problem 4.8 k-**Edge Minimization Problem** ($EdgeMin(G, k)$)

Given: (1) A network $G = \langle V, E \rangle$; (2) the connectivity function f as defined in Eq. (4.2); and (3) the budget k.

Output: A set with k edges $E' \subseteq E$, such that $C(G \setminus E')$ (i.e., the connectivity in $G \setminus E'$) is minimized.

We first prove that $EdgeMin(G, k)$ is NP-hard by constructing a polynomial reduction from $MaxCover(n, m, k)$ problem. Similar to the rationale in the previous proof, we assume that $1 < k < m$.

Given an instance of $MaxCover(n, m, k)$, we construct an m-edge star-shaped network G (i.e., all the m edges share one common endpoint). Specifically, each subset in $\mathcal{S}$ corresponds to an edge in G and each element e_i in $\mathcal{U}$ represents a valid subgraph G_i in the connectivity function. To construct subgraph G_i, we first locate all the subsets in $\mathcal{S}$ that contain element e_i, and map them into the corresponding edges in G. Then the sub-star formed by those edges can be viewed as a valid subgraph G_i with $f(G_i) = w_i$. The removal of any edge from G_i would destroy the completeness of the valid subgraph. Consequently, we have n valid subgraphs in G. Similarly, as the sets in $\mathcal{S}$ are distinct and not mutually exclusive, the resulting valid subgraphs are guaranteed to be non-independent. Therefore, the solution of $MaxCover(n, m, k)$ would be equivalent to the solution of $EdgeMin(G, k)$, which proves that $EdgeMin(G, k)$ is NP-hard.

Again, given a candidate solution set of k node, we can verify its optimality by comparing to the $\binom{m}{k}$ all possible solution sets of size k. As calculating the impact of each set take polynomial time w.r.t. the network size and $\binom{m}{k}$ is also polynomial to the size of the network, the overall complexity to verify the optimality of the candidate solution is also polynomial to the size of the network, which indicates that $EdgeMin(G, k)$ is in NP.

Therefore, the $EdgeMin(G, k)$ problem is NP-complete.   □

Figure 4.1 gives an illustration of the reduction from an instance of max k-coverage problem $MaxCover(n, m, k)$ to k-edge minimization problem $EdgeMin(G, k)$. Again, edges with multiple colors indicate their participation in multiple valid subgraphs.

Fundamental Limit #2: Approximability

Based on the hardness result of NETCOM, we further reveal the approximability of NETCOM in Theorem 4.9, which says that $(1 - 1/e)$ is indeed the best approximation ratio a polynomial algorithm can achieve unless $NP \subseteq DTIME(n^{O(\log \log n)})$.

Theorem 4.9   Approximability of NETCOM. $(1 - 1/e)$ *is the best approximation ratio for the* NETCOM *problem in polynomial time, unless* $NP \subseteq DTIME(n^{O(\log \log n)})$.

Proof. We prove this by contradiction. In the proof of Theorem 4.3, we show that max k-Coverage problem is polynomially reducible to the NETCOM problem, which implies that if there is an α-approximation algorithm that can solve NETCOM in polynomial time with $\alpha > (1 - 1/e)$, there will be an α-approximation algorithm for max k-Coverage as well. However, it was proved in [54] that the maximum k-coverage problem cannot be approximat-

with a factor better than $(1 - 1/e)$ unless $NP \subseteq DTIME(n^{O(\log \log n)})$, which contradicts with our assumption. Hence, we conclude that there is no polynomial algorithm for the NETCOM problem with an approximation ratio greater than $(1 - 1/e)$, unless $NP \subseteq DTIME(n^{O(\log \log n)})$.
□

Since the greedy strategy in Algorithm 4.8 guarantees a $(1 - 1/e)$ approximation ratio, Theorem 4.9 implies that the greedy algorithm is the best polynomial algorithm for NETCOM in terms of its approximation ratio unless $NP \subseteq DTIME(n^{O(\log \log n)})$.

4.1.3   PROPOSED ALGORITHM

In this section, we start with detailing the algorithmic challenges of the network connectivity optimization (NETCOM) problem and then present an effective algorithm, followed by some analysis in terms of its effectiveness and efficiency.

Algorithmic Challenges of NETCOM

In the greedy strategy (Algorithm 4.8), a key step is to calculate the impact score of each network element, i.e., $I(o) = C(G) - C(G \setminus \{o\})$ (step 4). As we mentioned in Chapter 2, many network connectivity measures $C(G)$ can be calculated or well approximated by a function of top-r eigenvalues of its adjacency matrix (i.e., $C(G) = F(\Lambda^{(r)})$, where $F()$ is the function of eigenvalues). Therefore, the core step of calculating $I(o)$ is to compute $\Lambda^{(r)}$ on $G \setminus \{o\}$, which takes $O(m)$ time (say using the classic Lanczos method). Consequently, simply recomputing $C(G \setminus \{o\})$ for each network element from scratch would make the entire algorithm $O(mn)$ for node-level optimization problems and $O(m^2)$ for edge-level optimization problems, neither of which is computationally feasible in large networks. To address this issue, existing literature often resorts to matrix perturbation theory. Its key idea is to view the deletion of a network element o as a perturbation to the original network (i.e., $\tilde{A} = A + \Delta A$). Thus, the new eigenvalues (and hence the new connectivity measure $C(G \setminus \{o\})$) can be approximated from the eigenvalues and eigenvectors of the original network in constant time, making the overall algorithm linear w.r.t. the size of the input network [18, 19]. However, for networks with small eigen-gaps, the approximation accuracy of matrix perturbation-based methods might deteriorate quickly, if not collapse at all. This issue might persist even if we switch to computationally more expensive high-order matrix perturbation theory [18, 19]. Thus, the main algorithmic challenge is how to accurately approximate the top-r eigenvalues of the input network after a node/edge deletion.

CONTAIN: The Proposed Algorithm

We propose a new updating algorithm for the top-r eigenvalues after node/edge deletion. In order to maintain the linear complexity of the entire greedy algorithm, we seek to update the top-r eigenvalues in *constant* time for each node/edge deletion operation.

Our key observation is as follows. In classic matrix perturbation theory (whether the first-order matrix perturbation theory or its high-order variants), a fundamental assumption is that

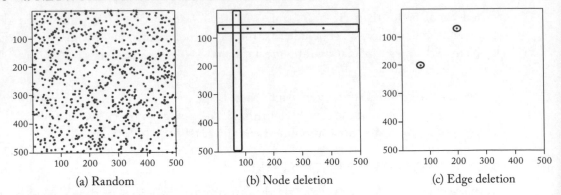

(a) Random   (b) Node deletion   (c) Edge deletion

Figure 4.2: Illustrations and comparison of random perturbation matrix (a), which is dense and potentially full-rank vs. perturbation matrices by node deletion (b) and edge deletion (c), both of which are sparse and low-rank.

the perturbation matrix $\Delta \mathbf{A}$ is a random matrix whose spectrum is well-bounded as illustrated in Figure 4.2a. However, such assumption does not hold in the node/edge deletion scenario (Figure 4.2b and c), in which the perturbation matrix $\Delta \mathbf{A}$ is *sparse* and *low-rank*. Armed with this observation, we propose an effective eigen-pair update algorithm for node/edge deletion based on partial-QR decomposition. Unlike matrix perturbation-based methods, which would inevitably introduce approximation error in the procedure, the proposed algorithm does not introduce any additional error when computing the impact score $I(o)$, and it runs in constant time for each node/edge operation.

The proposed CONTAIN algorithm is presented in Algorithm 4.9. Overall, it follows the greedy strategy (Algorithm 4.8). In detail, we first compute the top-r eigen-pairs of the network and compute the connectivity score of the original network (steps 2–3). From steps 4–19, we iteratively select the element with the highest impact score. When evaluating the impact of each valid element, we first construct the perturbation matrix $\Delta \mathbf{A}$ for the corresponding element and then perform eigen-decomposition on it (steps 6–7). Particularly, for node deletion operation suppose the removed node v has a set of neighbor nodes $\mathcal{N}_v$. Then the resulting perturbation matrix $\Delta \mathbf{A}$ has $\Delta \mathbf{A}(v, \mathcal{N}_v) = \Delta \mathbf{A}(\mathcal{N}_v, v) = -1$, which is a rank-2 sparse matrix. Therefore, $\mathbf{U}$ and $\mathbf{\Lambda}_\Delta$ can be directly expressed as an $n \times 2$ matrix and a 2×2 matrix, respectively. Moreover let $n_v = |\mathcal{N}_v|$, the non-zero entries in the eigenvector matrix of $\Delta \mathbf{A}$ are

$$\mathbf{U}_\Delta(v, 1) = \frac{1}{\sqrt{2}}, \quad \mathbf{U}_\Delta(v, 2) = \frac{1}{\sqrt{2}}$$

$$\mathbf{U}_\Delta(\mathcal{N}_v, 1) = -\frac{1}{\sqrt{2n_v}}, \quad \mathbf{U}_\Delta(\mathcal{N}_v, 2) = \frac{1}{\sqrt{2n_v}} \qquad (4.$$

Algorithm 4.9 The CONTAIN Algorithm

Input: (1) The adjacency matrix of the network $\mathbf{A}$; (2) the associated eigen-function $F()$ for connectivity $C(G)$; (3) rank r; (4) the network operation (node vs. edge deletion); and (5) a positive integer k

Output: A set of network elements $\mathcal{X}$ of size k

1: initialize $\mathcal{X}$ to be empty
2: compute $[\mathbf{U}^{(r)}, \mathbf{\Lambda}^{(r)}] \leftarrow$ top-r eigen-pairs of matrix $\mathbf{A}$
3: compute $C(G) \leftarrow F(\mathbf{\Lambda}^{(r)})$
4: **for** $i = 1$ to k **do**
5:     **for** each valid element o in G **do**
6:         $\Delta\mathbf{A} \leftarrow$ the perturbation matrix by element o's deletion
7:         $[\mathbf{U}_\Delta, \mathbf{\Lambda}_\Delta] \leftarrow$ eigen-pairs of $\Delta\mathbf{A}$
8:         $\mathbf{R} \leftarrow$ upper triangular matrix from $[\mathbf{U}^{(r)}, \mathbf{U}_\Delta]$'s partial-QR decomposition
9:         $\mathbf{\Lambda}_z \leftarrow$ eigenvalues of $\mathbf{Z} = \mathbf{R}[\mathbf{\Lambda}^{(r)}, \mathbf{0}; \mathbf{0}, \mathbf{\Lambda}_\Delta]\mathbf{R}'$
10:        compute $I(o) \leftarrow C(G) - F(\mathbf{\Lambda}_z)$
11:     **end for**
12:     add $\tilde{o} = \operatorname{argmax}_o I(o)$ to $\mathcal{X}$
13:     update $C(G) \leftarrow C(G) - I(\tilde{o})$ and set $I(\tilde{o}) \leftarrow -1$
14:     $\Delta\mathbf{A} \leftarrow$ the perturbation matrix by element $\tilde{o}$'s deletion
15:     $[\mathbf{U}_\Delta, \mathbf{\Lambda}_\Delta] \leftarrow$ eigen-pairs of $\Delta\mathbf{A}$
16:     $[\mathbf{Q}, \mathbf{R}] \leftarrow$ partial-QR decomposition of $[\mathbf{U}^{(r)}, \mathbf{U}_\Delta]$
17:     $[\mathbf{U}_z, \mathbf{\Lambda}_z] \leftarrow$ eigen-pairs of $\mathbf{Z} = \mathbf{R}[\mathbf{\Lambda}^{(r)}, \mathbf{0}; \mathbf{0}, \mathbf{\Lambda}_\Delta]\mathbf{R}'$
18:     update $\mathbf{U}^{(r)} \leftarrow (\mathbf{Q}\mathbf{U}_z)^{(r)}$, $\mathbf{\Lambda}^{(r)} \leftarrow \mathbf{\Lambda}_z^{(r)}$, $\mathbf{A} \leftarrow \mathbf{A} + \Delta\mathbf{A}$
19: **end for**
20: return $\mathcal{X}$

---

and the eigenvalue matrix of $\Delta\mathbf{A}$ is

$$\mathbf{\Lambda}_\Delta = \begin{bmatrix} \sqrt{n_v} & 0 \\ 0 & -\sqrt{n_v} \end{bmatrix}. \tag{4.4}$$

In the edge deletion scenario, the perturbation matrix $\Delta\mathbf{A}$ corresponding to the removal of edge $\langle u, v \rangle$ has only two non-zero entries $\Delta\mathbf{A}(u, v) = \Delta\mathbf{A}(v, u) = -1$ and $u \neq v$, which is also a rank-2 matrix. Then, the only non-zero entries in $\mathbf{U}_\Delta$ are

$$\mathbf{U}_\Delta(u, 1) = \frac{1}{\sqrt{2}}, \quad \mathbf{U}_\Delta(u, 2) = \frac{1}{\sqrt{2}}$$

$$\mathbf{U}_\Delta(v, 1) = -\frac{1}{\sqrt{2}}, \quad \mathbf{U}_\Delta(v, 2) = \frac{1}{\sqrt{2}} \tag{4.5}$$

and the eigenvalue matrix $\mathbf{\Lambda}_\Delta$ is

$$\mathbf{\Lambda}_\Delta = \begin{bmatrix} 1 & 0 \\ 0 & -1 \end{bmatrix}. \tag{4.6}$$

With the eigenvector matrix of $\Delta\mathbf{A}$, we proceed to perform partial-QR decomposition on $[\mathbf{U}^{(r)}, \mathbf{U}_\Delta]$ in step 8. As $\mathbf{U}^{(r)}$ is already orthonormal, the $\mathbf{Q}$ matrix in the decomposition can be written as the concatenation of $\mathbf{U}^{(r)}$ and two orthogonal vectors in unit length as follows:

$$\mathbf{Q} = [\mathbf{U}^{(r)}, \frac{\mathbf{q}_1}{\|\mathbf{q}_1\|}, \frac{\mathbf{q}_2}{\|\mathbf{q}_2\|}]. \tag{4.7}$$

By the Gram–Schmidt process, we have

$$\begin{aligned} \mathbf{q}_1 &= \mathbf{U}_\Delta(:, 1) - \mathbf{U}^{(r)}\mathbf{r}_1 \\ \mathbf{q}_2 &= \mathbf{U}_\Delta(:, 2) - \mathbf{U}^{(r)}\mathbf{r}_2 + \mathbf{r}_1'\mathbf{r}_2\frac{\mathbf{q}_1}{\|\mathbf{q}_1\|^2}, \end{aligned} \tag{4.8}$$

where $\mathbf{r}_1 = \mathbf{U}^{(r)'}\mathbf{U}_\Delta(:, 1)$ and $\mathbf{r}_2 = \mathbf{U}^{(r)'}\mathbf{U}_\Delta(:, 2)$.

For node-level operations, we have

$$\mathbf{r}_1 = \mathbf{U}^{(r)'}\mathbf{U}_\Delta(:, 1) = \frac{1}{\sqrt{2}} \left(\mathbf{U}^{(r)}(v, :) - \frac{1}{\sqrt{n_v}} \sum_{u \in \mathcal{N}_v} \mathbf{U}^{(r)}(u, :) \right)'$$

$$\mathbf{r}_2 = \mathbf{U}^{(r)'}\mathbf{U}_\Delta(:, 2) = \frac{1}{\sqrt{2}} \left(\mathbf{U}^{(r)}(v, :) + \frac{1}{\sqrt{n_v}} \sum_{u \in \mathcal{N}_v} \mathbf{U}^{(r)}(u, :) \right)'. \tag{4.9}$$

While for edge-level operations, we have

$$\mathbf{r}_1 = \mathbf{U}^{(r)'}\mathbf{U}_\Delta(:, 1) = \frac{1}{\sqrt{2}}(\mathbf{U}^{(r)}(u, :) - \mathbf{U}^{(r)}(v, :))'$$

$$\mathbf{r}_2 = \mathbf{U}^{(r)'}\mathbf{U}_\Delta(:, 2) = \frac{1}{\sqrt{2}}(\mathbf{U}^{(r)}(u, :) + \mathbf{U}^{(r)}(v, :))'. \tag{4.10}$$

Correspondingly, the upper-triangular matrix $\mathbf{R}$ can be written as

$$\mathbf{R} = \begin{bmatrix} \mathbf{I} & \mathbf{r}_1 & \mathbf{r}_2 \\ 0 & \|\mathbf{q}_1\| & -\dfrac{\mathbf{r}_1'\mathbf{r}_2}{\|\mathbf{q}_1\|} \\ 0 & 0 & \|\mathbf{q}_2\| \end{bmatrix}. \tag{4.11}$$

By the definition of $\mathbf{q}_1, \mathbf{q}_2$ in Eq. (4.8) together with the orthonormal property of the eigenvectors, the norms of $\mathbf{q}_1$ and $\mathbf{q}_2$ can be computed indirectly with two $r \times 1$ vectors $\mathbf{r}_1$ and $\mathbf{r}$ as

$$\begin{aligned} \|\mathbf{q}_1\| &= \sqrt{1 - \|\mathbf{r}_1\|^2} \\ \|\mathbf{q}_2\| &= \sqrt{1 - \|\mathbf{r}_2\|^2 - \frac{(\mathbf{r}_1'\mathbf{r}_2)^2}{1 - \|\mathbf{r}_1\|^2}}. \end{aligned} \tag{4.1}$$

This enables us to compute $\|\mathbf{q}_1\|$ and $\|\mathbf{q}_2\|$ without explicitly constructing $\mathbf{q}_1$ and $\mathbf{q}_2$, which reduces the cost of step 8 from $O(nr)$ to $O(r)$. It can be proved that by setting $\mathbf{Z} = \mathbf{R}[\mathbf{\Lambda}^{(r)}, \mathbf{0}; \mathbf{0}, \mathbf{\Lambda}_\Delta]\mathbf{R}'$, the eigenvalues of $\mathbf{Z}$ are just the top eigenvalues of the perturbed matrix $\mathbf{A} + \Delta\mathbf{A}$, and the top eigenvectors of $\mathbf{A} + \Delta\mathbf{A}$ can be calculated by $\mathbf{QU}_z$ (step 18). Therefore, we only need $\mathbf{\Lambda}_z$ to compute the impact score of element o (step 10). After scanning all the valid elements in the current network, we choose the one with the largest impact score and add it to the element set $\mathcal{X}$ (steps 12–13). Then, we update the network and its eigen-pairs (steps 14–18). The procedure to update eign-pairs is similar to that of computing the impact score for a given network element (steps 6–9), with the following subtle difference. In order to just compute the impact score of a given network element, we only need the updated eigenvalues. This is crucial as it saves the computation of (1) constructing $\mathbf{q}_1$ and $\mathbf{q}_2$, (2) finding the eigenvectors of $\mathbf{Z}$, and (3) updating the eigenvectors of the perturbed matrix $\mathbf{A} + \Delta\mathbf{A}$, which in turn helps maintain constant time complexity for each inner for-loop (steps 5–11).

Algorithm 4.9 can be easily extended to address the local connectivity minimization problem by properly approximating the local connectivity impact score at step 10. It is worth noting that for some complex local connectivity measures like local natural connectivity or local length-t path capacity, it is often time-consuming to directly calculate local connectivity $C_\mathcal{T}(G)$ and element impact $I_\mathcal{T}(o)$ on $C_\mathcal{T}(G)$ than the global connectivity $C(G)$ and $I(o)$. This is mainly because that $C_\mathcal{T}(G)$ and $I_\mathcal{T}(o)$ cannot be directly calculated with the eigen-pairs of the network. To efficiently address this problem, we propose the following heuristic to measure $C_\mathcal{T}(G)$ and $I_\mathcal{T}(o)$.

Lemma 4.10   Local Impact Computation. *Let $C(G \setminus \mathcal{T})$ and $I^{G \setminus \mathcal{T}}(o)$ denotes the global connectivity of graph $G \setminus \mathcal{T}$ and impact of element o on $C(G \setminus \mathcal{T})$, then $C_\mathcal{T}(G) = C(G) - C(G \setminus \mathcal{T})$, and $I_\mathcal{T}(o) = I(o) - I^{G \setminus \mathcal{T}}(o)$.*

Proof. The first equation naturally holds by the definition of connectivity measures. Here we proceed to prove the second part. By the definition of $I_\mathcal{T}(o)$, we have

$$I_\mathcal{T}(o) = C_\mathcal{T}(G) - C_\mathcal{T}(G \setminus \{o\}).$$

By the fact that $C_\mathcal{T}(G) = C(G) - C(G \setminus \mathcal{T})$, the above equation can be re-write as

$$\begin{aligned} I_\mathcal{T}(o) &= (C(G) - C(G \setminus \mathcal{T})) - (C(G \setminus \{o\}) - C(G \setminus \{o\} \cup \mathcal{T}) && (4.13) \\ &= (C(G) - C(G \setminus \{o\})) - (C(G \setminus \mathcal{T}) - C(G \setminus \{o\} \cup \mathcal{T}) \\ &= I(o) - I^{G \setminus \mathcal{T}}(o). \end{aligned}$$

$\square$

Proof and Analysis

In this section, we analyze the proposed CONTAIN algorithm w.r.t. its effectiveness and efficiency.

Effectiveness. The effectiveness of CONTAIN is summarized in Lemma 4.11, which says that the computation of the impact score for each valid network element in the inner for-loop does not introduce any extra approximation error.

Lemma 4.11 **Effectiveness of CONTAIN.** *Suppose* $\mathbf{A}$ *is approximated with its top-r eigen-pairs with error* $\mathbf{E}$ *(i.e.,* $\mathbf{A} = \mathbf{U}^{(r)}\mathbf{\Lambda}^{(r)}\mathbf{U}^{(r)'} + \mathbf{E}$*), then the* $\mathbf{\Lambda}_z$ *and* $\mathbf{QU}_z$ *returned in Algorithm 4.9 can be used to approximate* $\tilde{\mathbf{A}}$ *as its top eigen-pairs with no extra error.*

Proof. As $\mathbf{A} = \mathbf{U}^{(r)}\mathbf{\Lambda}^{(r)}\mathbf{U}^{(r)'} + \mathbf{E}$ and $\Delta\mathbf{A} = \mathbf{U}_\Delta\mathbf{\Lambda}_\Delta\mathbf{U}'_\Delta$, then $\tilde{\mathbf{A}}$ can be expressed as

$$\tilde{\mathbf{A}} = \mathbf{U}^{(r)}\mathbf{\Lambda}^{(r)}\mathbf{U}^{(r)'} + \mathbf{U}_\Delta\mathbf{\Lambda}_\Delta\mathbf{U}'_\Delta + \mathbf{E}$$
$$= [\mathbf{U}^{(r)}, \mathbf{U}_\Delta] \begin{bmatrix} \mathbf{\Lambda}^{(r)} & 0 \\ 0 & \mathbf{\Lambda}_\Delta \end{bmatrix} [\mathbf{U}^{(r)}, \mathbf{U}_\Delta]' + \mathbf{E}. \qquad (4.14)$$

Perform partial-QR decomposition on $[\mathbf{U}^{(r)}, \mathbf{U}_\Delta]$ as $[\mathbf{U}^{(r)}, \mathbf{U}_\Delta] = \mathbf{QR}$, we get orthonormal basis for $\tilde{\mathbf{A}}$ and an upper triangular matrix $\mathbf{R}$. Then the perturbed matrix $\tilde{\mathbf{A}}$ can be rewritten as

$$\tilde{\mathbf{A}} = \mathbf{QR} \begin{bmatrix} \mathbf{\Lambda}^{(r)} & 0 \\ 0 & \mathbf{\Lambda}_\Delta \end{bmatrix} \mathbf{R}'\mathbf{Q}' + \mathbf{E}. \qquad (4.15)$$

Let $\mathbf{Z} = \mathbf{R}[\mathbf{\Lambda}^{(r)}, 0; 0, \mathbf{\Lambda}_\Delta]\mathbf{R}'$ and perform eigen-decomposition on $\mathbf{Z}$ as $\mathbf{Z} = \mathbf{U}_z\mathbf{\Lambda}_z\mathbf{U}'_z$, $\tilde{\mathbf{A}}$ is now equivalent to

$$\tilde{\mathbf{A}} = \mathbf{QU}_z\mathbf{\Lambda}_z\mathbf{U}'_z\mathbf{Q}' + \mathbf{E} = (\mathbf{QU}_z)\mathbf{\Lambda}_z(\mathbf{QU}_z)' + \mathbf{E}. \qquad (4.16)$$

Since both $\mathbf{Q}$ and $\mathbf{U}_z$ are orthonormal, we have $(\mathbf{QU}_z)(\mathbf{QU}_z)' = \mathbf{I}$. Thus, $\mathbf{\Lambda}_z$ and $\mathbf{QU}_z$ can be viewed as the top eigen-pairs of $\tilde{\mathbf{A}}$. As the approximation error remains to be $\mathbf{E}$ in Eq. (4.16), it implies that no extra error is introduced in the procedure, which completes the proof. $\square$

As the eigenvalues in real networks are often skewed [36], the above impact scores can be approximated with top-r eigenvalues. Analysis in existing literature ([125] and [14]) shows that the truncated approximations for triangle capacity and natural connectivity can achieve high accuracy with only top-50 eigenvalues, which enables a great acceleration on impact score approximation.

Efficiency. The complexity of the proposed CONTAIN algorithm is summarized in Lemma 4.12, which says it is linear in both time and space.

Lemma 4.12 **Complexity of CONTAIN.** *The time complexity of CONTAIN for node-level connectivity optimization is* $O(k(mr + nr^3))$. *The time complexity of CONTAIN for edge-level connectivity optimization is* $O(k(mr^3 + nr^2))$. *The space complexity of CONTAIN is* $O(nr + m$

Proof. In the CONTAIN algorithm, computing top-r eigen-pairs and connectivity $C(G)$ would take $O(nr^2 + mr)$ and $O(r)$, respectively. To compute the impact score for each node/edge (steps 5–11), it takes $O(d_v r)$ (d_v is the degree of node v) for node v, and $O(r)$ for each edge to get the upper triangular matrix $\mathbf{R}$ in step 8. Since performing eigendecomposition on $\mathbf{Z}$ at step 9 takes $O(r^3)$, the complexity to collect impact scores for all the nodes/edges are $O(nr^3 + mr)$ and $O(mr^3)$, respectively. Picking out the node/edge with the highest impact score in the current iteration would cost $O(n)$ for node-level operations and $O(m)$ for edge-level operations. At the end of the iteration, updating the eigen-pairs of the network takes the complexity of $O(nr^2 + r^3)$. As we have $r \ll n$, the overall time complexity to select k nodes would be $O(k(mr + nr^3))$; and the complexity to select k edges would be $O(k(mr^3 + nr^2))$.

For space complexity, it takes $O(n + m)$ to store the entire network, $O(nr)$ to calculate and store the top-r eigen-pair of the network, $O(n)$ to store the impact scores for all the nodes in node level optimization scenarios and $O(m)$ to store the impact scores for all the edges, the eigen-pair update requires a space of $O(nr)$. Therefore, the overall space complexity for CONTAIN is $O(nr + m)$. $\qquad\square$

CONTAIN+: The Closed-Form Heuristics

Here we provide the heuristics for the triangle capacity and natural connectivity optimization problems, which can be easily extended to other similar connectivity measures.

Impact Approximation. For the triangle capacity optimization problem, the impact of a node/edge is the number of triangles that the node/edge participates in, which can be directly approximated with the eigen-pairs of the current network.

Lemma 4.13   Closed-Form Impact Score for Triangle Capacity. *Given a network G with adjacency matrix $\mathbf{A}$ and eigen-pair $(\mathbf{U}, \mathbf{\Lambda})$. The number of triangles that node v participates in is $I(v) = \sum_{i=1}^{n} \frac{\lambda_i^3 \mathbf{u}_i^2(v)}{2}$; the number of triangles that edge $\langle u, v \rangle$ participates in is $I(\langle u, v \rangle) = \sum_{i=1}^{n} \lambda_i^2 \mathbf{u}_i(u) \mathbf{u}_i(v)$.*

Proof. The first part of the lemma has been proved in [125]. Here we proceed to prove the second part.

The number of triangles that edge $\langle u, v \rangle$ involves in equals to the number of length-2 paths from node u to node v, which equals $\mathbf{A}^2(u, v)$. As $\mathbf{A} = \mathbf{U\Lambda U}'$, we have $\mathbf{A}^2 = \mathbf{U\Lambda}^2 \mathbf{U}'$. Therefore, $\mathbf{A}^2(u, v) = \sum_{i=1}^{n} \lambda_i^2 \mathbf{u}_i(u) \mathbf{u}_i(v)$. $\qquad\square$

Lemma 4.14   Closed-Form Impact Score for Natural Connectivity. *Given a network G with adjacency matrix $\mathbf{A}$ and eigen-pair $(\mathbf{U}, \mathbf{\Lambda})$. The impact of node v on natural connectivity is $I(v) = \sum_{i=1}^{n} e^{\lambda_i} \mathbf{u}_i^2(v)$; the impact of edge $\langle u, v \rangle$ is $I(\langle u, v \rangle) = \sum_{i=1}^{n} \frac{e^{\lambda_i} - 1}{\lambda_i} \mathbf{u}_i(u) \mathbf{u}_i(v)$.*

Proof. Natural connectivity can be viewed as an aggregation of weighted closed-walks [51]. For node v, its impact on the number of length-t closed-walks is proportional to $\mathbf{A}^t(v, v)$, so its overall impact on natural connectivity can be expressed as $I(v) = \sum_{j=0}^{\infty} \frac{\mathbf{A}^j(v,v)}{j!}$. As we have $\mathbf{A} = \mathbf{U \Lambda U}'$ and $\mathbf{A}^j(v, v) = \sum_{i=1}^n \lambda_i^j \mathbf{u}_i^2(v)$, we have

$$I(v) = \sum_{j=0}^{\infty} \frac{1}{j!} \sum_{i=1}^n \lambda_i^j \mathbf{u}_i^2(v) = \sum_{i=1}^n \sum_{j=1}^{\infty} \frac{1}{j!} \lambda_i^j \mathbf{u}_i^2(v) \tag{4.17}$$
$$= \sum_{i=1}^n e^{\lambda_i} \mathbf{u}_i^2(v).$$

For edge $\langle u, v \rangle$, the number of length-t closed-walks it participates in equals to the number of length-$(t-1)$ walks from node u to node v, which can be expressed as $\mathbf{A}^{t-1}(u, v)$. Therefore, its overall impact on natural connectivity can be written as $I(\langle u, v \rangle) = \sum_{j=1}^{\infty} \frac{\mathbf{A}^{j-1}(u,v)}{j!}$. Let $\mathbf{T} = \sum_{j=1}^{\infty} \frac{\mathbf{A}^{j-1}}{j!}$; then we have

$$\mathbf{AT} = \sum_{j=1}^{\infty} \frac{\mathbf{A}^j}{j!} = \sum_{j=0}^{\infty} \frac{\mathbf{A}^j}{j!} - \mathbf{I} = e^{\mathbf{A}} - \mathbf{I}.$$

Based on the above equation, we have

$$\mathbf{T} = \mathbf{A}^{-1}(e^{\mathbf{A}} - \mathbf{I}) = \mathbf{U \Lambda}^{-1} \mathbf{U}'(\mathbf{U}(e^{\mathbf{A}} - \mathbf{I})\mathbf{U}') \tag{4.18}$$
$$= \mathbf{U \Lambda}^{-1}(e^{\mathbf{A}} - \mathbf{I})\mathbf{U}'.$$

Thus, $I(\langle u, v \rangle) = \mathbf{T}(u, v) = \sum_{i=1}^n \frac{e^{\lambda_i} - 1}{\lambda_i} \mathbf{u}_i(u) \mathbf{u}_i(v)$. $\qquad \square$

Effectiveness of CONTAIN+. As we have mentioned, the impact score of a node/edge is often approximated with the top-r eigen-pairs of the network. Let $(\mathbf{U}, \mathbf{\Lambda})$ be the eigen-pairs of network G, $(\tilde{\mathbf{U}}, \tilde{\mathbf{\Lambda}})$ be the eigen-pairs of network $G \setminus \{v\}$. To estimate the impact of node v on the connectivity of the network, CONTAIN needs to utilize the eigenvalues from both the original network (i.e., $\mathbf{\Lambda}$) and the perturbed network (i.e., $\tilde{\mathbf{\Lambda}}$) for the calculation (Algorithm 4.9, step 10), while CONTAIN+ only relies on the eigen-pairs in the original network (i.e., $(\mathbf{U}, \mathbf{\Lambda})$). Take triangle capacity optimization under node operations as an example. The impact of node v can be approximated as $I(v)_{\text{CONTAIN}} = \sum_{i=1}^r \frac{\lambda_i^3}{6} - \sum_{i=1}^r \frac{\tilde{\lambda}_i^3}{6}$ by CONTAIN; or it can be approximated as $I(v)_{\text{CONTAIN+}} = \sum_{i=1}^r \frac{\lambda_i^3 \mathbf{u}_i^2(v)}{2}$ by CONTAIN. Suppose the exact impact of node v is $I(v)_{Exact}$, then we can define the approximation error of CONTAIN as $err_{\text{CONTAIN}} = I(v)_{Exact} - I(v)_{\text{CONTAIN}}$ and the error of CONTAIN+ $err_{\text{CONTAIN+}} = I(v)_{Exact} - I(v)_{\text{CONTAIN+}}$. In Lemma 4.15, we give the analysis on err_{CONTAIN} and $err_{\text{CONTAIN+}}$.

Lemma 4.15 *The error of* CONTAIN *for triangle capacity impact approximation is* $err_{CONTAIN} = \frac{\sum_{i=r+1}^{n} \lambda_i^3 - \tilde{\lambda}_i^3}{6}$; *and the approximation error for* CONTAIN+ *is* $err_{CONTAIN+} = \sum_{i=r+1}^{n} \frac{\lambda_i^3 u_i^2(v)}{2}$.

Proof. As we have $I(v)_{Exact} = \sum_{i=1}^{n} \frac{\lambda_i^3 u_i^2(v)}{2} = \frac{\sum_{i=1}^{n} \lambda_i^3 - \tilde{\lambda}_i^3}{6}$, Lemma 4.15 naturally holds when $I(v)_{CONTAIN}$ and $I(v)_{CONTAIN+}$ are subtracted from $I(v)_{Exact}$, respectively.   □

Lemma 4.15 implies that when the removed node has small effect on the bottom-$(n - r)$ eigenvalues of the underlying network, CONTAIN is preferred as $err_{CONTAIN}$ would be small. While in networks with very skewed eigenvalue distributions, CONTAIN+ is preferred as the bottom-$(n - r)$ eigenvalues are small in magnitude. Similar analysis can be derived for edge-level operation scenarios and natural connectivity optimization scenarios, which is omitted for brevity.

Complexity of CONTAIN+. In Theorem 4.16, we give the time and space complexity of CONTAIN+.

Theorem 4.16 *The time complexity of* CONTAIN+ *for node level connectivity optimization is* $O(nr^2 + mr + knr^2)$. *The time complexity for edge level connectivity optimization is* $O(k(mr + nr^2))$. *The space complexity of* CONTAIN+ *is* $O(nr + m)$.

Proof. The CONTAIN+ algorithm is generally similar to the CONTAIN algorithm except for the node/edge impact calculation part. Therefore, CONTAIN+ also need to take $O(nr^2 + mr)$ to compute the top-r eigen-pairs. In each iteration, it takes CONTAIN+ $O(nr)$ time to get all the impact scores for each node or $O(mr)$ for the scores of each edge. After picking out the highest impact node/edge with $O(n)$ or $O(m)$ complexity, we update the eigen-pairs of the network with the method used in CONTAIN with $O(nr^2 + r^3)$ complexity. Therefore, the overall complexity for node-level connectivity minimization is $O(nr^2 + mr + knr^2)$ and the complexity for edge-level connectivity minimization is $O(nr^2 + mr + k(mr + nr^2))$, which can be simplified as $O(k(mr + nr^2))$.

As for the space complexity, CONTAIN+ also needs space to store all the top-r eigen-pairs and the impact scores for nodes/edges. Since no extra computation space is introduced in CONTAIN+ compared to CONTAIN, the overall space complexity for CONTAIN+ is still $O(nr + m)$.   □

4.1.4 EXPERIMENTAL EVALUATION

In this section, we evaluate the proposed CONTAIN algorithm. All experiments are designed to answer the following two questions:

Table 4.2: Statistics of datasets

Domain	Dataset	Number of Nodes	Number of Edges	Average Degree
Infrastructure	Airport	2,833	7,602	5.37
	Oregon	5,296	10,097	3.81
Biology	Chemical	6,026	69,109	22.94
	Disease	4,256	30,551	14.36
	Gene	7,604	14,071	3.70
Collaboration	Astrph	18,772	198,050	21.10
	Hepth	9,877	25,985	5.26
	Aminer	1,211,749	4,756,194	7.85
Social	Eucore	1,005	16,064	31.97
	Fb	4,039	88,234	43.69
	youtube	1,138,499	2,990,443	5.25

- **Effectiveness.** How effective is the proposed CONTAIN algorithm in minimizing various connectivity measures?

- **Efficiency.** How efficient and scalable is the proposed CONTAIN algorithm?

Experiment Setup

Datasets. We perform experiments on 10 different datasets from 4 different domains, including: AIRPORT: an air traffic network that represents the direct flight connections between internal U.S. airports;[4] OREGON: an autonomous system network which depicts the information transferring relationship between routers from [65]; CHEMICAL: a network based on [30] that shows the similarity between different chemicals; DISEASE: a network that depicts the similarity between different diseases [30]; GENE: a protein-protein interaction network based on [30]; ASTRPH: a collaboration network between authors whose papers were submitted to Astro Physics category on Arxiv [66]; HEPTH: a collaboration network between authors whose papers were submitted to High Energy Physics (Theory category) on Arxiv [65]; AMINER: a collaboration network between researchers in the Aminer datasets [117]; EUCORE: the email correspondence network from a large European research institution [66]; FB: a social circle network collected from Facebook [82]; and YOUTUBE: a friendship network among youtube users [120]. The statistics of those datasets are listed in Table 4.2.

Comparing Methods. We compare the proposed algorithm with the following methods: (1) *Degree*: selecting top-k nodes (edges) with the largest degrees; specifically, for edge $\langle u, v \rangle$,

---

[4]http://www.levmuchnik.net/Content/Networks/NetworkData.html

d_u and d_v denote the degrees for its endpoints, respectively, the score for $\langle u, v \rangle$ is $\min\{d_u, d_v\}$.[5]
(2) *PageRank*: selecting top-k nodes (edges) with the largest PageRank scores [94] (the corresponding edge score is the minimum PageRank score among its two endpoints). (3) *Eigenvector*: selecting top-k nodes (edges) with the largest eigenvector centrality scores [90] (the corresponding edge score is the minimum eigenvector centrality score among its endpoints).
(4) *Netshield/Netmelt*: selecting top-k nodes (edges) that minimize the leading eigenvalue of the network [18, 19]. (5) *MIOBI*: a greedy algorithm that employs first-order matrix perturbation method to estimate element impact score and update eigen-pairs [14]. (6) *MIOBI-S*: a variant of MIOBI that selects top-k nodes (edges) in one batch without updating the eigen-pairs of the network. (7) *MIOBI-H*: a variant of MIOBI that employs high-order matrix perturbation method to update eigen-pairs [17]. (8) *Exact*: a greedy algorithm that recomputes the top-r eigen-pairs to estimate the impact score for each candidate node/edge. For the results reported here, we set rank $r = 80$ for all the top-r eigen-pairs based approximation methods (methods (5)–(8) and the proposed CONTAIN method).

Evaluation Metrics. The performance of the algorithm is evaluated by the impact of its selected elements $I(\mathcal{X}) = C(G) - C(G \setminus \mathcal{X})$. The larger the $I(\mathcal{X})$ is, the more effective the algorithm is. For a given dataset, connectivity measure and network operation, we normalize $I(\mathcal{X})$ by that of the best method, so that the results across different datasets are comparable in the same plot.

Machine and Repeatability. All the experiments in this work are performed on a machine with two processors (Intel Xeon 3.5 GHz) with 256 GB of RAM. The algorithms are programmed with MATLAB using a single thread.

Effectiveness
Effectiveness of CONTAIN and CONTAIN+. We compare the proposed algorithm and the baseline methods on three connectivity measures (leading eigenvalue, number of triangles, and natural connectivity) by both node-level operations and edge-level operations on all datasets in our experiment. Since the *Exact* method needs to recompute the top-r eigen-pairs for each candidate node/edge which is very time-consuming, its results would be absent on some large datasets (e.g., AMINIER and YOUTUBE) where it does not finish the computation within 24 hours. In our experiment, the budget for node-level operations is $k = 20$, the budget for edge-level operations is $k = 200$. The results are shown from Figures 4.4–4.9. We can see that the proposed CONTAIN (the red solid bar) and CONTAIN+ (the red hollow bar) (1) are very close to the *Exact* method (the black hollow bar) and (2) consistently outperforms all the other alternative methods. In the meanwhile, the proposed CONTAIN and CONTAIN+ algorithms are much faster than *Exact*, as will be shown later.

To study the effectiveness of CONTAIN and CONTAIN+ for the local connectivity minimization problem, we experiment on the CHEMICAL data and compare the performance of

---

[5]We use $\min\{d_u, d_v\}$ as edge score to ensure that both ends of the top-ranked edges are high degree nodes.

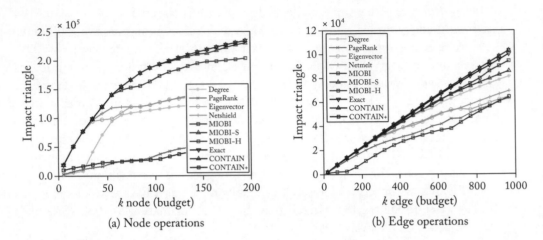

Figure 4.3: The optimization results on the number of local triangles on the CHEMICAL dataset.

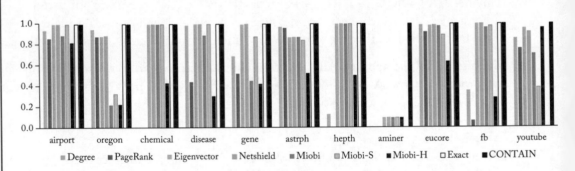

Figure 4.4: The optimization results on leading eigenvalue with node-level operations.

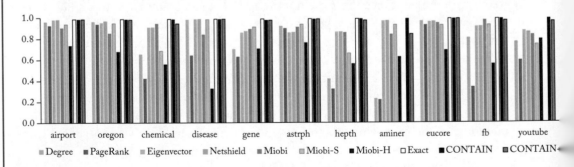

Figure 4.5: The optimization results on the number of triangles with node-level operations.

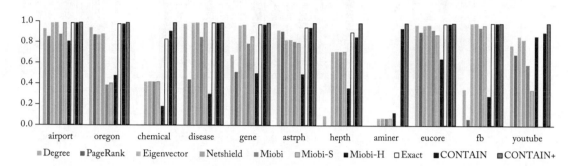

Figure 4.6: The optimization results on natural connectivity with node-level operations.

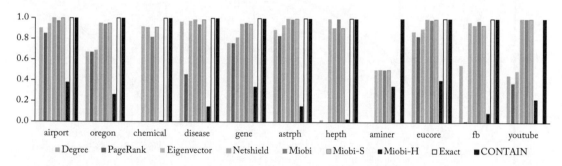

Figure 4.7: The optimization results on leading eigenvalue with edge-level operations.

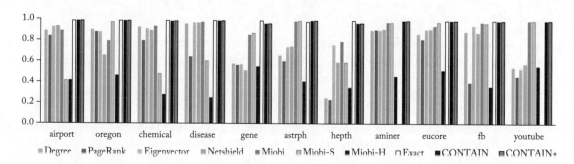

Figure 4.8: The optimization results on the number of triangles with edge-level operations.

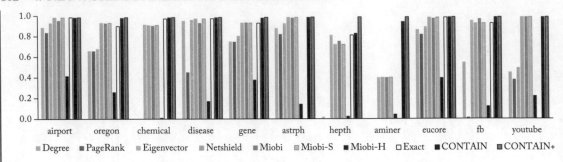

Figure 4.9: The optimization results on natural connectivity with edge-level operations.

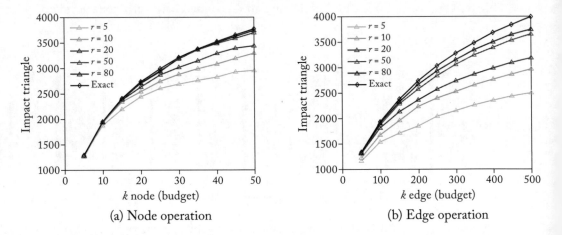

(a) Node operation

(b) Edge operation

Figure 4.10: The effect of r on optimizing the number of triangles on the CHEMICAL dataset.

different methods for minimizing the local triangle capacity in the network. From Figure 4.3, it is obvious to see that both CONTAIN and CONTAIN+ can achieve similar performance with the *Exact* algorithm and outperform all other heuristic methods.

Effect of Rank r. The main parameter that affects the performance of CONTAIN is the rank r. To study the effect of r, we change r from 5–80 to minimize the number of triangles on the CHEMICAL dataset and compare them with the *Exact* method. The results are shown in Figure 4.10. From Figure 4.10, it is obvious to see that as r increases, the performance of CONTAIN increases accordingly, which is consistent with our effectiveness analysis. With $r = 80$, the performance of CONTAIN is very close to the *Exact* method with different k.

Table 4.3: Time complexity comparison for node operations

	Degree	PageRank	Eigenvector	NetShield	MIOBI	Exact	CONTAIN
Time Complexity	$O(m + nk)$	$O(m + nk)$	$O(m + nk)$	$O(m + nk^2)$	$O(k(mr + nr^2))$	$O(k(n^2r^2 + nmr))$	$O(k(mr + nr^3))$

Table 4.4: Time complexity comparison for edge operations

	Degree	PageRank	Eigenvector	NetMelt	MIOBI	Exact	CONTAIN
Time Complexity	$O(mk)$	$O(mk)$	$O(mk)$	$O(n + mk)$	$O(k(mr + nr^2))$	$O(k(nmr^2 + m^2r))$	$O(k(mr^3 + nr^2))$

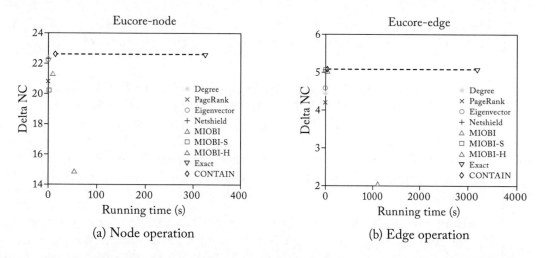

(a) Node operation   (b) Edge operation

Figure 4.11: The quality vs. running time trade-off on Eucore. The budget for node operations is $k = 20$, the budget for edge operations is $k = 200$.

Efficiency

Efficiency of CONTAIN. In Tables 4.3 and 4.4, we have compared the time complexity of all the methods in its base versions and we assume the budge $k < \log n$ for node operations and $k < \log m$ for edge operations.

With that, we present the quality vs. running time trade-off of different methods for optimizing the natural connectivity (the most complicated connectivity measure) on the Eucore dataset in Figure 4.11. In both node-level and edge-level optimization scenarios, the proposed CONTAIN achieves very similar performance as *Exact*. In terms of the running time, CONTAIN is orders of magnitude faster than *Exact*. Although the running time of other baseline methods is similar to CONTAIN, their performance (y-axis) is not as good as CONTAIN.

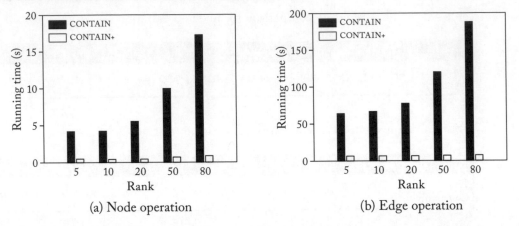

(a) Node operation   (b) Edge operation

Figure 4.12: The running time comparison between CONTAIN and CONTAIN+ on the CHEMICAL dataset. The budget for both node and edge operations is $k = 20$.

Efficiency of CONTAIN+. To justify the efficiency of CONTAIN+, we compare the running time of CONTAIN and CONTAIN+ on the CHEMICAL dataset in Figure 4.12. We can see that the running time of CONTAIN+ is orders of magnitudes faster than the CONTAIN algorithm. Moreover, as rank r increases, the running time of CONTAIN would increase in polynomial order due to the eigendecomposition operation for node/edge impact score approximation; while the running time of CONTAIN+ only shows a slightly linear increase across different rank settings.

Scalability of CONTAIN. The scalability results of CONTAIN are presented in Figure 4.13. As we can see, the proposed CONTAIN algorithm scales linearly w.r.t. the size of the input network (i.e., both the number of nodes and edges), which is consistent with Lemma 4.12.

4.2 CONNECTIVITY OPTIMIZATION IN MULTI-LAYERED NETWORKS

The multi-layered networks are fundamentally different from the single-layered networks due to the cross-layer dependencies between different networks. Such dependency has made multi-layered networks are more vulnerable to external attacks because their nodes can be affected by both within-layer connections and cross-layer dependencies. That is, even a small disturbance in one layer/network may be amplified in all its dependent networks through cross-layer dependencies, and cause cascade failure to the entire system. For example, when the supporting facilities (e.g., power stations) in a metropolitan area are destroyed by natural disasters like hurricanes or earthquakes, the resulting blackout would not only put tens of thousands of people in dark for a long time, but also paralyze the telecom network and cause a great interruption of

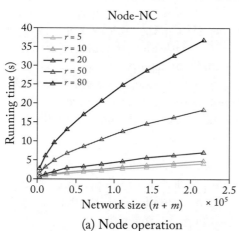

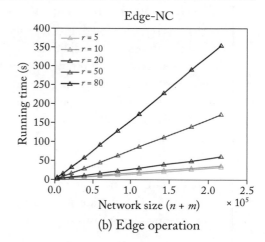

(a) Node operation

(b) Edge operation

Figure 4.13: The scalability of CONTAIN. The budget for both node and edge operations is $k = 20$.

the transportation network. Therefore, it is of key importance to identify crucial nodes in the supporting layer/network, whose loss would lead to a catastrophic failure of the entire system, so that the countermeasures can be taken proactively. On the other hand, accessibility issues extensively exist in multi-layered network mining tasks. To manipulate the connectivity in layers with limited accessibility, one can only operate via the nodes from accessible layers that have large impact to target layers. Taking the multi-layered network depicted in Figure 2.3a for example, assume that the only accessible layer in the system is the control layer and the goal is to minimize the connectivity in the satellite communication layer and physical layer simultaneously under k attacks, the only strategy we could adopt is to select a set of k nodes from the control layer, whose failure would cause the largest reduction on the connectivity of the two target layers.

To tackle the connectivity optimization[6] problem in multi-layered networks, great efforts have been made from different research areas for manipulating *two-layered* interdependent network systems [12, 41, 96, 108, 110]. In [32], De Domenico et al. proposed a method to identify versatile nodes in multi-layered networks by evaluating their eigenvector centrality and PageRank centrality. The selected versatile nodes are fundamentally different from our high-impact nodes for network connectivity in three aspects. First, is that their centrality measures cannot capture the collective impact of a node set on the network. Second, is that our proposed network connectivity is directly related to only within-layer links, while cross-layer dependency is the trigger for connectivity changes. The two types of links should be treated differently rather than mixed up for a unified centrality calculation. Last, the globally crucial nodes in the entire

---

[6]In this work, the connectivity optimization problem is defined as minimizing the connectivity of a target layer by removing a fixed number of nodes in the control layer.

system may not be able to provide an optimal solution to minimize the connectivity in specific target layer(s). Although much progress has been made, the optimization strategies used in two-layered networks may still be sub-optimal or even misleading in multi-layered network settings, where we want to simultaneously optimize the connectivity in multiple layers by manipulating one common supporting layer. On the theoretical side, the optimality of the connectivity optimization problem of generic multi-layered networks is largely unknown.

This work aims to address the above challenges, and the main contributions can be summarized as follows.

- *Connectivity Optimization.* We show that for *any* network connectivity measures in the SuB-LiNE family, the connectivity optimization problem with the proposed MuLaN model enjoys the diminishing returns property, which naturally lends itself to a family of provable near-optimal optimization algorithms with linear complexity.

- *Empirical Evaluations.* We perform extensive experiments based on real data sets to validate the effectiveness and efficiency of the proposed algorithms.

4.2.1   PROBLEM DEFINITION

In this section, we start with the main symbols used in this work (Table 4.5) and then give the formal definition of the multi-layered network connectivity optimization problem. We use bold uppercase letters for matrices (e.g., $\mathbf{A}$, $\mathbf{B}$), bold lowercase letters for column vectors (e.g., $\mathbf{a}$, $\mathbf{b}$) and calligraphic font for sets (e.g., $\mathcal{A}$, $\mathcal{B}$). The transpose of a matrix is denoted with a prime, i.e., $\mathbf{A}'$ is the transpose of matrix $\mathbf{A}$.

With the above symbols[7] and the definition of multi-layered networks in Chapter 2, we formally define the connectivity optimization problem (OPERA) on the proposed MuLaN model for multi-layered networks as follows.

Problem 4.17   Opera on MuLaN

Given: (1) A multi-layered network $\Gamma =< \mathbf{G}, \mathcal{A}, \mathcal{D}, \theta, \varphi >$; (2) a control layer $\mathbf{A}_l$; (3) an impac function $\mathbb{I}(.)$; and (4) an integer k as operation budget.

Output: A set of k nodes $\mathcal{S}_l$ from the control layer ($\mathbf{A}_l$) such that $\mathbb{I}(\mathcal{S}_l)$ (the overall impact o $\mathcal{S}_l$) is maximized.

In the above definition, the control layer $\mathbf{A}_l$ indicates the sources of the "attack;" an the $g \times 1$ vector $\boldsymbol{\alpha}$ indicates the target layer(s) as well as their relative weights. For instance, i Figure 2.3a, we can choose layer-1 as the control layer (indicated by the strike sign); and s $\boldsymbol{\alpha} = [0\ 1\ 0\ 1\ ]'$, which means that both layer-2 and layer-4 are the target layers (indicated by th

---

[7]In this section, we use the adjacency matrix $\mathbf{A}_i$ to represent the corresponding network. Thus, the $\mathbf{A}_i$ here is equivale to the $\mathbf{G}_i$ in Section 2.

Table 4.5: Main symbols for MuLaN

Symbol	Definition and Description
$\mathbf{A}, \mathbf{B}$	the adjacency matrices (bold uppercase)
$\mathbf{a}, \mathbf{b}$	column vectors (bold lowercase)
$\mathcal{A}, \mathcal{B}$	sets (calligraphic)
$\mathbf{A}(i,j)$	the element at ith row jth column in matrix $\mathbf{A}$
$\mathbf{A}(i,:)$	the ith row of matrix $\mathbf{A}$
$\mathbf{A}(:,j)$	the jth column of matrix $\mathbf{A}$
$\mathbf{A}'$	transpose of matrix $\mathbf{A}$
$\mathbf{G}$	the layer-layer dependency matrix
$\mathcal{A}$	adjacency matrices at each layers $\mathcal{A} = \{\mathbf{A}_1, ..., \mathbf{A}_g\}$
$\mathcal{D}$	cross-layer node-node dependency matrices
θ, φ	one to one mapping functions
Γ	multi-layered network MuLaN $\Gamma = <\mathbf{G}, \mathcal{A}, \mathcal{D}, \theta, \varphi>$
$\mathcal{S}_i, \mathcal{T}_i, ...$	node sets in layer $\mathbf{A}_i$ (calligraphic)
$\mathcal{S}_{i \rightarrow j}$	nodes in $\mathbf{A}_j$ that depend on nodes $\mathcal{S}$ in $\mathbf{A}_i$
$\mathcal{N}(\mathcal{S}_i)$	nodes and cross-layer links that depend on $\mathcal{S}_i$
m_i, n_i	number of edges and nodes in layer $\mathbf{A}_i$
$\lambda_{<\mathbf{A},j>}, \mathbf{u}_{<\mathbf{A},j>}$	jth largest eigenvalue (in module) and eigenvector $\mathbf{A}$
$\lambda_{\mathbf{A}}, \mathbf{u}_{\mathbf{A}}$	first eigenvalue and eigenvector of network $\mathbf{A}$
$C(\Lambda_i, f)$	connectivity of network $\mathbf{A}_i$ under mapping function f
$I_{\mathbf{A}}(\mathcal{S}_i)$	impact of node set $\mathcal{S}_i$ on network $\mathbf{A}$
$\mathbb{I}(\mathcal{S}_i)$	overall impact of node set $\mathcal{S}_i$ on MuLaN

star signs) with equal weights between them. In this example, once a subset of nodes $\mathcal{S}$ in layer-1 are attacked/deleted (e.g., shaded circle nodes), all the nodes from layer-2 and layer-3 that are dependent on $\mathcal{S}$ (e.g., shaded parallelogram and triangle nodes) will be disabled/deleted, which will in turn cause the disfunction of the nodes in layer 4 (e.g., shaded diamond nodes) that depend on the affected nodes in layer-2 or layer-3. Our goal is to choose k nodes from layer-1 that have the maximal impact on both layer-2 and layer-4, i.e., to simultaneously decrease the connectivity $C(\mathbf{A}_2, f)$ and $C(\mathbf{A}_4, f)$ as much as possible.

4.2.2   THEORETICAL ANALYSIS

In this section, we present the major theoretical results of the connectivity optimization prob lem (OPERA) on multi-layered networks defined in Problem 4.1. It says that for **any** connectivit function $C(\mathbf{A}, f)$ in the SubLine family (Eq. (2.1)), for **any** multi-layered network in the Mu LaN family (Definition (1)), the connectivity optimization problem (OPERA, Problem 4.1) bear **diminishing returns property**.

Let us start with the base case, where there is only one single input network. In this case $\Gamma = <\mathbf{G}, \mathcal{A}, \mathcal{D}, \theta, \varphi>$ in Problem 4.1 degenerates to a single-layered network $\mathbf{A}$, which is bot the control layer as well as the sole control target (i.e., $\alpha = 1$, and $l = 1$). With such a setting Lemma 4.18 says that OPERA enjoys the *diminishing returns property*, that is, the overall impac function $\mathbb{I}(\mathcal{S}_1)$ (which in this case degenerates to $I(\mathcal{S})$, i.e., the impact of the node set $\mathcal{S}$ o network G itself) is (a) monotonically non-decreasing; (b) sub-modular; and (c) normalized.

Lemma 4.18   Diminishing Returns Property of a Single-Layered Network. *Given a simp undirected, unweighted network* $\mathbf{A}$, *for any connectivity function* $C(\mathbf{A}, f)$ *in the* SubLine *famil the impact function* $I(\mathcal{S})$ *is (a) monotonically non-decreasing; (b) sub-modular; and (c) normalize where* $\mathcal{S} \subseteq \mathbf{A}$.

Proof. By the definition of the connectivity function $C(\mathbf{A}, f)$ (Eq. (2.1)), we have

$$I(\mathcal{S}) = \sum_{\pi \subseteq \mathbf{A}} f(\pi) - \sum_{\pi \subseteq \mathbf{A} \setminus \mathcal{S}} f(\pi) = \sum_{\pi \subseteq \mathbf{A}, \, \pi \cap \mathcal{S} \neq \Phi} f(\pi),$$

where Φ is the empty set. Apparently, we have $I(\Phi) = 0$ since $f(\Phi) = 0$. In other words, th impact function $I(\mathcal{S})$ is normalized.

Let $\mathcal{I}, \mathcal{J}, \mathcal{K}$ be three sets and $\mathcal{I} \subseteq \mathcal{J}$. We further define three sets as follows: $\mathcal{S} = \mathcal{I}$ $\mathcal{K}$, $\mathcal{T} = \mathcal{J} \cup \mathcal{K}$, $\mathcal{R} = \mathcal{J} \setminus \mathcal{I}$.

We have

$$I(\mathcal{J}) - I(\mathcal{I}) = \sum_{\pi \subseteq \mathbf{A}, \, \pi \cap \mathcal{J} \neq \Phi} f(\pi) - \sum_{\pi \subseteq \mathbf{A}, \, \pi \cap \mathcal{I} \neq \Phi} f(\pi)$$

$$= \sum_{\pi \subseteq \mathbf{A}, \, \pi \cap (\mathcal{J} \setminus \mathcal{I}) \neq \Phi} f(\pi) = \sum_{\pi \subseteq \mathbf{A}, \, \pi \cap \mathcal{R} \neq \Phi} f(\pi)$$

$$\geq 0$$

which proves the monotonicity of the impact function $I(\mathcal{S})$.

Let us define another set $\mathcal{P} = \mathcal{T} \setminus \mathcal{S}$. We have that $\mathcal{P} = (\mathcal{J} \cup \mathcal{K}) \setminus (\mathcal{I} \cup \mathcal{K}) = \mathcal{R} \setminus (\mathcal{R}$ $\mathcal{K}) \subseteq \mathcal{R} = \mathcal{J} \setminus \mathcal{I}$. Then, we have

$$I(\mathcal{T}) - I(\mathcal{S}) = \sum_{\pi \subseteq \mathbf{A}, \, \pi \cap \mathcal{P} \neq \Phi} f(\pi) \leq I(\mathcal{J}) - I(\mathcal{I})$$

which completes the proof of the sub-modularity of the impact function $I(\mathcal{S})$. $\qquad\qquad\square$

In order to generalize Lemma 4.18 to an arbitrary, generic member in the MuLaN family, we first need the following lemma, which says that the set-ordering relationship in a supporting layer is preserved through dependency links in all dependent layers of MuLaN.

Lemma 4.19   Set-Ordering Preservation Lemma on DAG. *Given a multi-layered network* $\Gamma = <\mathbf{G}, \mathcal{A}, \mathcal{D}, \theta, \varphi>$, *and the dependency network* $\mathbf{G}$ *is a directed acyclic graph (DAG). For two node sets* $\mathcal{I}_l, \mathcal{J}_l$ *in* $\mathbf{A}_l$ *such that* $\mathcal{I}_l \subseteq \mathcal{J}_l$, *we have that in any layer* $\mathbf{A}_i$ *in the system,* $\mathcal{I}_{l \to i} \subseteq \mathcal{J}_{l \to i}$ *holds, where* $\mathcal{I}_{l \to i}$ *and* $\mathcal{J}_{l \to i}$ *are the node sets in layer* $\mathbf{A}_i$ *that depend on* $\mathcal{I}_l$ *and* $\mathcal{J}_l$ *in layer* $\mathbf{A}_l$, *respectively.*

Proof. If $l = i$, we have $\mathcal{J}_{l \to i} = \mathcal{J} \subseteq \mathcal{I}_{\updownarrow \to)} = \mathcal{I}$ and Lemma 4.19 holds.

Second, if layer-i does not depend on layer-l either directly or indirectly, we have $\mathcal{J}_{l \to i} = \mathcal{I}_{l \to i} = \Phi$, where Φ is an empty set. Lemma 4.19 also holds.

If layer-i does depend on layer-l through the layer-layer dependency network $\mathbf{G}$, we will prove Lemma 4.19 by induction. Let $\text{len}(l \rightsquigarrow i)$ be the maximum length of the path from node l to node i on the layer-layer dependency network $\mathbf{G}$. Since $\mathbf{G}$ is a DAG, we have that $\text{len}(l \rightsquigarrow i)$ is a finite number.

Base Case. Suppose $\text{len}(l \rightsquigarrow i) = 1$, we have that layer-$i$ directly depends on layer-l. Let $\mathcal{R}_l = \mathcal{J}_l \setminus \mathcal{I}_l$. We have that

$$\mathcal{J}_{l \to i} = \mathcal{I}_{l \to i} \cup \mathcal{R}_{l \to i} \supseteq \mathcal{I}_{l \to i} \tag{4.19}$$

which complete the proof for the base case where $\text{len}(l \rightsquigarrow i) = 1$.

Induction Step. Suppose Lemma 4.19 holds for $\text{len}(l \rightsquigarrow i) \leq q$, where q is a positive integer. We will prove that Lemma 4.19 also holds for $\text{len}(l \rightsquigarrow i) = q + 1$.

Suppose layer-i directly depends on layer-i_x ($x = 1, ..., d(i)$, where $d(i)$ is the in-degree of node i on $\mathbf{G}$). Since $\mathbf{G}$ is a DAG, we have that $\text{len}(l \rightsquigarrow i_x) \leq q$. By the induction hypothesis, given $\mathcal{I}_l \subseteq \mathcal{J}_l$, we have that $\mathcal{I}_{l \to i_x} \subseteq \mathcal{J}_{l \to i_x}$.

We further have $\mathcal{I}_{l \to i} = \cup_{x=1,...,d(i)} (\mathcal{I}_{l \to i_x})_{i_x \to i}$.

Let $\mathcal{R}_{l \to i_x} = \mathcal{J}_{l \to i_x} \setminus \mathcal{I}_{l \to i_x}$ for $x = 1, ..., d(i)$. We have that

$$\mathcal{J}_{l \to i} = [\cup_{x=1,...,d(i)} (\mathcal{I}_{l \to i_x})_{i_x \to i}] \tag{4.20}$$
$$\cup [\cup_{x=1,...,d(i)} (\mathcal{R}_{l \to i_x})_{i_x \to i}]$$
$$= \mathcal{I}_{l \to i} \cup \mathcal{R}_{l \to i} \supseteq \mathcal{I}_{l \to i}$$

which completes the proof of the induction step.

Putting everything together, we have completed the proof for Lemma 4.19. $\qquad\square$

Notice that in the proof of Lemma 4.19, it requires the layer-layer dependency network $\mathbf{G}$ to be a DAG so that the longest path from the control layer $\mathbf{A}_l$ to any target layer $\mathbf{A}_t$ is of finite length. To further generalize it to arbitrary dependency structures, we need the following lemma, which says that the dependent paths from the control layer to the layer in any arbitrarily structured dependency network can be reduced to a DAG.

Lemma 4.20   **DAG Dependency Reduction Lemma.** *Given a multi-layered network* $\Gamma =< \mathbf{G}, \mathcal{A}, \mathcal{D}, \theta, \varphi >$ *with arbitrarily structured layer-layer dependency network* $\mathbf{G}$, *a control layer* $\mathbf{A}_l$, *and a target layer* $\mathbf{A}_t$, *the dependent paths constructed by Algorithm 4.10 can be reduced to a DAG.*

Proof. In Algorithm 4.10, the Tarjan Algorithm is first used to find out all strongly connected components $\mathcal{V} = \{SC_1, SC_2, \ldots, SC_f\}$ in layer-layer dependency network $\mathbf{G}$. The cross-component dependency edges are denoted as $\mathcal{E} = \{E_{i,j}\}_{i,j=1,\ldots,f, i \neq j}$ where $\langle u, v \rangle \in E_{i,j}$ iff $\mathbf{G}(u, v) = 1$ and $\mathbf{A}_u \in SC_i$, $\mathbf{A}_v \in SC_j$. Based on the node set $\mathcal{V}$ and the edge set $\mathcal{E}$, a directed meta-graph $\mathcal{G}$ can be constructed where $\mathcal{G}(u, v) = 1$ iff $E_{i,j} \neq \phi$. The meta-graph $\mathcal{G}$ is acyclic. Otherwise, the cycle in $\mathcal{G}$ would be merged into a large strongly connected component by Tarjan Algorithm in the first place. Suppose the control layer $\mathbf{A}_l$ and the target layer $\mathbf{A}_t$ are located in strongly connected component SC_i and SC_j, respectively, then a set of acyclic paths $\mathcal{P}$ from SC_i and SC_j can be extracted from $\mathcal{G}$. To show that the dependent paths from $\mathbf{A}_l$ to $\mathbf{A}_t$ is DAG, we only need to show that each meta-path $\mathbf{P} \in \mathcal{P}$ can be unfolded into a DAG.

Here we proceed to show how a meta-path $\mathbf{P}$ can be represented with a DAG. As the nodes in $\mathbf{P}$ are strongly connected components that contain cycles, and the edges in $\mathbf{P}$ contain corresponding cross-component edges that would not introduce any cycles, representing $\mathbf{P}$ with a DAG can be converted to a problem of unfolding the cyclic dependent paths in a strongly connected component into a DAG. As described in Algorithm 4.13, a strongly connected component $\mathcal{Q}$ can be partitioned into two parts: (1) a DAG that contains all acyclic links (denoted as $\mathbf{G}_{\mathcal{Q},0}$) and (2) links that enclose cycles in $\mathcal{Q}$ (denoted as $E_{\mathcal{Q},0}$). Therefore, given a strongly connected component $\mathcal{Q}$ and a set of dependent nodes $\{\mathcal{T}_v\}_{\mathbf{A}_v \in \mathbf{Q}}$ in $\mathcal{Q}$, the dependent cycle can be replaced by a chain of $\mathbf{G}_{\mathcal{Q},0}$'s replicas, where the two adjacent replicas are linked by $E_{\mathcal{Q},}$ until the number of the dependent nodes in the connected component converges (steps 5–2 in Algorithm 4.12). As the number of dependent nodes keeps increasing in each iteration and is upper bounded by the total number of nodes in $\mathcal{Q}$, the repetition is guaranteed to stop at stable state within finite iterations. Since $\mathbf{G}_{\mathcal{Q},0}$ is a DAG, the links ($E_{\mathcal{Q},0}$) between each replica $\{\mathbf{G}_{\mathcal{Q},1}, \ldots, \mathbf{G}_{\mathcal{Q},L}\}$ would not introduce any cycle, the resulting graph $\mathbf{G}_{\mathcal{Q}}$ is also a DAG. Therefore, the dependent paths constructed by Algorithm 4.10 from $\mathbf{A}_l$ and $\mathbf{A}_t$ can be represented a DAG.

A complete DAG reduction algorithm is summarized from Algorithms 4.10–4.13.

---

[8]A widely used strongly connect component detection algorithm in [121].

Algorithm 4.10 DAG Reduction Algorithm

Input: (1) A multi-layered network Γ, (2) a control layer $\mathbf{A}_l$, (3) a set of node $\mathcal{S}_l$ in layer $\mathbf{A}_l$, and (4) a target layer $\mathbf{A}_t$

Output: (1) A DAG $\mathbf{G}_D$ that contains all the dependent paths from $\mathcal{S}_l$ in layer $\mathbf{A}_l$ to $\mathbf{A}_t$ and (2) $\mathcal{S}_{l \to t}$

1: find out all strongly connected components in $\mathbf{G}$ as $\mathcal{V} \leftarrow \{\mathcal{SC}_1, \mathcal{SC}_2, \dots, \mathcal{SC}_f\}$ with Tarjan Algorithm[8]
2: set $\mathcal{E} \leftarrow \{E_{i,j}\}_{i,j=1,\dots,f}$, where $\langle u, v \rangle \in E_{i,j}$ iff $\mathbf{G}(u, v) = 1$ and $\mathbf{A}_u \in \mathcal{SC}_i$, $\mathbf{A}_v \in \mathcal{SC}_j$
3: construct meta-graph $\mathcal{G}$ from $\mathcal{V}$ s.t. $\mathcal{G}(i, j) = 1$ iff $E_{i,j} \neq \phi$
4: $\mathcal{SC}_i \leftarrow$ connected component that contains $\mathbf{A}_l$
5: $\mathcal{SC}_j \leftarrow$ connected component that contains $\mathbf{A}_t$
6: find out all paths $\mathcal{P}$ from $\mathcal{SC}_i$ to $\mathcal{SC}_j$ in $\mathcal{G}$
7: initialize $\mathbf{G}_D \leftarrow \phi, \mathcal{S}_{l \to t} = \phi$
8: **for** each path $\mathbf{P}$ in $\mathcal{P}$ **do**
9: $\quad [\mathbf{G}_D^{\mathbf{P}}, \mathcal{S}_{l \to t}^{\mathbf{P}}] \leftarrow unfoldPath(\mathbf{P}, \mathcal{G}, \mathcal{S}_l, \Gamma, \mathcal{V}, \mathcal{E})$
10: $\quad \mathbf{G}_D \leftarrow \mathbf{G}_D \cup \mathbf{G}_D^{\mathbf{P}}, \mathcal{S}_{l \to t} \leftarrow \mathcal{S}_{l \to t} \cup \mathcal{S}_{l \to t}^{\mathbf{P}}$
11: **end for**
12: return $[\mathbf{G}_D, \mathcal{S}_{l \to t}]$

In Algorithm 4.10, step 1 runs Tarjan Algorithm [121] to find out all the strongly connected components in layer-layer dependency network $\mathbf{G}$. Step 2 collects all cross-component edges into set $\mathcal{E}$. In the following step, a meta-graph $\mathcal{G}$ is constructed based on $\mathcal{V}$ and $\mathcal{E}$. In steps 4 and 5, the connected components that contain control layer and target layer are located ($\mathcal{SC}_i$ and $\mathcal{SC}_j$). Step 6 finds out all meta-paths from $\mathcal{SC}_i$ to $\mathcal{SC}_j$. In step 7, the final DAG $\mathbf{G}_D$ and dependent node set $\mathcal{S}_{l \to t}$ are initialized as empty sets. From steps 8–11, the DAG $\mathbf{G}_D^{\mathbf{P}}$ and dependent node set $\mathcal{S}_{l \to t}^{\mathbf{P}}$ for each path $\mathbf{P}$ in $\mathcal{P}$ are calculated by function $unfoldPath$, and are used to update $\mathbf{G}_D$ and $\mathcal{S}_{l \to t}$ in step 10.

To illustrate how Algorithm 4.10 works, we present a simple example in Figure 4.14. In the example, the dependency network $\mathbf{G}$ contains three layers, where $\mathbf{A}_1$ is the control layer and $\mathbf{A}_3$ is the target layer. Specifically, $\mathbf{A}_2$ is a dependent layer of $\mathbf{A}_1$, while $\mathbf{A}_2$ and $\mathbf{A}_3$ are interdependent to each other. The toy example has two strongly connected components $\{\mathcal{SC}_1, \mathcal{SC}_2\}$ and one cross-component edge set $E_{1,2} = \{\langle 1, 2 \rangle\}$. The meta-graph $\mathcal{G}$ is a link graph with just two nodes.

In Algorithm 4.11, the first connected component $\mathcal{Q}$ is initialized as the connected component that contains control layer $\mathbf{A}_l$ in step 2, the dependent nodes are initialized as $\mathcal{S}_l$ from steps 5–8 and the root layer $\mathbf{R}$ is initialized as the control layer $\mathbf{A}_l$. From steps 10–36, the DAG $\mathbf{G}_\mathcal{Q}^{\mathbf{P}}$ and the final dependent nodes in $\mathcal{Q}$ are calculated by the function $unfoldSC$ in step 11; $\mathbf{G}_\mathcal{Q}^{\mathbf{P}}$ is then added to the final DAG $\mathbf{G}_D^{\mathbf{P}}$ via cross-component links E_{i_q'', i_q} from steps 15–17. The

Algorithm 4.11 UnfoldPath: Construct DAG from Meta-Path

Input: (1) A meta-path $\mathbf{P} = \mathcal{SC}_i \to \dots \to \mathcal{SC}_j$, (2) a meta-graph $\mathcal{G}$, (3) a set of nodes $\mathcal{S}_l$ in $\mathbf{A}_l \in \mathcal{SC}_i$, (4) multi-layered network Γ, (5) all strongly connected components $\mathcal{V}$, and (6) all cross-component edges $\mathcal{E}$

Output: (1) A DAG $\mathbf{G}_D^{\mathbf{P}}$ and (2) $\mathcal{S}_{l \to t}^{\mathbf{P}}$

1: append ϕ to the end of meta-path $\mathbf{P}$
2: set $\mathcal{Q} = \mathcal{SC}_i$
3: $i_q \leftarrow$ index of connected component $\mathcal{Q}$ in meta-graph $\mathcal{G}$
4: $i_q'' \leftarrow -1$
5: **for** each layer $\mathbf{A}_v$ in $\mathcal{Q}$ **do**
6:     initialize $\mathcal{T}_v \leftarrow \phi$
7: **end for**
8: $\mathcal{T}_l \leftarrow \mathcal{T}_l \cup \mathcal{S}_l$
9: set root $\mathbf{R} \leftarrow \mathbf{A}_l$
10: **while true do**
11:     $[\mathbf{G}_{\mathcal{Q}}^{\mathbf{P}}, \{\mathcal{S}_{l \to v}^{\mathbf{P}}\}_{\mathbf{A}_v \in \mathcal{Q}}] \leftarrow unfoldSC(\mathcal{Q}, \{\mathcal{T}_v\}_{\mathbf{A}_v \in \mathcal{Q}}, \mathbf{R})$
12:     **if** $i_q'' = -1$ **then**
13:         $\mathbf{G}_D^{\mathbf{P}} \leftarrow \mathbf{G}_{\mathcal{Q}}^{\mathbf{P}}$
14:     **else**
15:         **for** each $\langle u, v \rangle \in E_{i_q'', i_q}$ **do**
16:             link layer $\mathbf{A}_u^{L_{i_q''}} \in \mathbf{G}_D^{\mathbf{P}}$ to layers $\{\mathbf{A}_v^x\}_{x=1,\dots,L_{i_q}} \in \mathbf{G}_{\mathcal{Q}}^{\mathbf{P}}$
17:         **end for**
18:     **end if**
19:     $\mathcal{Q}' \leftarrow \mathcal{Q}.successor()$
20:     **if** $\mathcal{Q}' = \phi$ **then**
21:         **break**
22:     **else**
23:         $i_q' \leftarrow$ index of $\mathcal{Q}'$ in meta-graph $\mathcal{G}$
24:         **for** each layer $\mathbf{A}_v$ in $\mathcal{Q}'$ **do**
25:             initialize $\mathcal{T}_v \leftarrow \phi$
26:         **end for**
27:         **for** each edge $\langle u, v \rangle \in E_{i_q, i_q'}$ **do**
28:             $\mathcal{T}_v \leftarrow \mathcal{T}_v \cup (\mathcal{S}_{l \to u}^{\mathbf{P}})_{u \to v}$
29:         **end for**
30:         $\mathbf{R} \leftarrow \mathbf{A}_r$, where $\mathbf{A}_r$ is a randomly picked layer from $\mathcal{Q}'$ with $\mathcal{T}_r \neq \phi$
31:         $\mathcal{Q} \leftarrow \mathcal{Q}'$
32:         $i_q'' \leftarrow i_q$
33:         $i_q \leftarrow i_q'$
34:     **end if**
35: **end while**
36: **return** $\mathbf{G}_D^{\mathbf{P}}, \mathcal{S}_{l \to t}^{\mathbf{P}}$

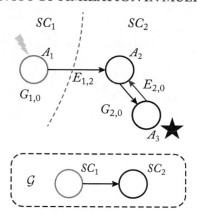

Figure 4.14: A cyclic dependency multi-layered network.

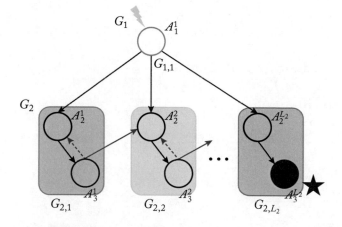

Figure 4.15: Constructed DAG for Figure 4.14.

initial dependent nodes for the next connected component $SC_{i'_q}$ are computed through cross-component links E_{i_q,i'_q} from steps 27–29. Step 30 is used to pick a root layer with non-empty dependent node set for $SC_{i'_q}$.

Algorithm 4.12 is used to unfold a strongly connected component into a DAG. In step 1, the input connected component Q is partitioned into a DAG $\mathbf{G}_{Q,0}$ and a set of cycle links $E_{Q,0}$. In step 2, the DAG $\mathbf{G}_Q$ is initialized by $\mathbf{G}_{Q,1}$, which is a replica of $\mathbf{G}_{Q,0}$. From steps 5–23, the algorithm keeps appending replicas of $\mathbf{G}_{Q,0}$ ($\mathbf{G}_{Q,c+1}$) onto $\mathbf{G}_Q$ (steps 8–16) until no new nodes are added to the dependent node set $\{\mathcal{T}_v\}_{\mathbf{A}_v \in Q}$ (steps 17–19).

For the example in Figure 4.14, SC_1 is unfolded as $\mathbf{G}_1$ with one node $\mathbf{A}_1^1$ in Figure 4.15. The initial dependent node set $\mathcal{T}_2$ for layer $\mathbf{A}_2$ can be calculated through $E_{1,2}$ as $\mathcal{T}_{1\to2}$. For SC_2, it is first partitioned into a DAG $\mathbf{G}_{2,0}$ and a cycle edge set $E_{2,0} = \{\langle \mathbf{A}_3, \mathbf{A}_2 \rangle\}$ as shown in

---

Algorithm 4.12 UnfoldSC: Construct DAG from Strongly Connected Component

---

Input: (1) A strongly connected component $\mathcal{Q}$, (2) a set of initial nodes for each layer $\{\mathcal{T}_v\}_{\mathbf{A}_v \in \mathbf{Q}}$, and (3) a root layer $\mathbf{R}$

Output: (1) A DAG $\mathbf{G}_\mathcal{Q}$ and (2) $\{\mathcal{S}_{l \to v}\}_{\mathbf{A}_v \in \mathcal{Q}}$

1: extract DAG and cycle edges $[\mathbf{G}_{\mathcal{Q},0}, E_{\mathcal{Q},0}] \leftarrow extractDAG(\mathcal{Q}, \mathbf{R})$
2: set $\mathbf{G}_{\mathcal{Q},1} \leftarrow \mathbf{G}_{\mathcal{Q},0}$, denote the layers in $\mathbf{G}_{\mathcal{Q},1}$ as $\{\mathbf{A}_v^1\}$
3: set $c \leftarrow 1$
4: initialize $\mathbf{G}_\mathcal{Q} \leftarrow \mathbf{G}_{\mathcal{Q},1}$
5: **while true do**
6:    $\{\mathcal{T}_v^c\}_{\mathbf{A}_v \in \mathcal{Q}} \leftarrow$ dependents of $\{\mathcal{T}_v\}_{\mathbf{A}_v \in \mathcal{Q}}$ in $\mathbf{G}_{\mathcal{Q},c}$
7:    update $\{\mathcal{T}_v\}_{\mathbf{A}_v \in \mathcal{Q}} \leftarrow \{\mathcal{T}_v \cup \mathcal{T}_v^c\}_{\mathbf{A}_v \in \mathcal{Q}}$
8:    set $\mathbf{G}_{\mathcal{Q},c+1} \leftarrow \mathbf{G}_{\mathcal{Q},0}$, layers in $\mathbf{G}_{\mathcal{Q},c+1}$ are denoted as $\{\mathbf{A}_v^{c+1}\}$
9:    extend $\mathbf{G}_\mathcal{Q} \leftarrow \mathbf{G}_\mathcal{Q} \cup \mathbf{G}_{\mathcal{Q},c+1}$
10:    **for** each edge $\langle u, v \rangle \in E_{\mathcal{Q},0}$ **do**
11:       $\mathcal{T}_{u \to v} \leftarrow$ all dependents of $\mathcal{T}_u$ in layer $\mathbf{A}_v$
12:       **if** $\mathcal{T}_{u \to v} \nsubseteq \mathcal{T}_v$ **then**
13:          add edge $\langle \mathbf{A}_u^c, \mathbf{A}_u^{c+1} \rangle$ to $\mathbf{G}_\mathcal{Q}$
14:          update $\mathcal{T}_v \leftarrow \mathcal{T}_v \cup \mathcal{T}_{u \to v}$
15:       **end if**
16:    **end for**
17:    **if** no edge added between $\mathbf{G}_{\mathcal{Q},c}$ and $\mathbf{G}_{\mathcal{Q},c+1}$ **then**
18:       remove $\mathbf{G}_{\mathcal{Q},c+1}$ from $\mathbf{G}_\mathcal{Q}$
19:       **break**
20:    **else**
21:       $c \leftarrow c + 1$
22:    **end if**
23: **end while**
24: return $[\mathbf{G}_\mathcal{Q}, \{\mathcal{T}_v\}_{\mathbf{A}_v \in \mathcal{Q}}]$

Figure 4.14. Suppose that the dependent node set in $\mathcal{SC}_2$ converges in L_2 iterations, then the DAG for $\mathcal{SC}_2$ can be presented with L_2 replicas of $\mathbf{G}_{2,0}$ linked by edges $\{\langle \mathbf{A}_3^c, \mathbf{A}_2^{c+1} \rangle\}_{c=1,\dots,L_2-}$ as shown in Figure 4.15. Putting it all together, the final DAG $\mathbf{G}_D$ can be constructed by linking $\mathbf{A}_1^1$ in $\mathbf{G}_1$ with $\{\mathbf{A}_2^x\}_{x=1,\dots,L_2}$ in $\mathbf{G}_2$.

Algorithm 4.13 is used to partition a strongly connected component $\mathcal{Q}$ into a DAG $\mathbf{G}$ and an edge set $E_{\mathcal{Q},0}$ that contains all cycle edges. The basic idea is to use Breadth-First-Search algorithm to traverse all the edges in the graph. In steps 1 and 2, $\mathbf{G}_{\mathcal{Q},0}$ and $E_{\mathcal{Q},0}$ are initialized as $\mathcal{Q}$ and ϕ, respectively. For each edge $\langle \mathbf{A}_u, \mathbf{A}_v \rangle$ in $\mathcal{Q}$, if $\mathbf{A}_v$ appears in $\mathbf{A}_u$'s ancestor list $\mathcal{L}$ then $\langle \mathbf{A}_u, \mathbf{A}_v \rangle$ would be removed from $\mathbf{G}_{\mathcal{Q},0}$ and added to $E_{\mathcal{Q},0}$ (steps 11–13).

Algorithm 4.13 ExtractDAG: Extract DAG from Strongly Connected Component

---

Input: (1) A strongly connected component Q and (2) a root layer $\mathbf{R}$ in the connected component

Output: (1) A DAG $\mathbf{G}_{Q,0}$ and (2) edge set $E_{Q,0}$ that contains all cycle edges

1: initialized $\mathbf{G}_{Q,0} \leftarrow Q$
2: initialized $E_{Q,0} \leftarrow \phi$
3: **for** each layer $\mathbf{A}_v \in Q$ **do**
4:    initialize its ancestor list $\mathcal{L}_v \leftarrow \phi$
5: **end for**
6: initialize a queue $\mathcal{T} \leftarrow \phi$
7: $\mathcal{T}.enqueue(\mathbf{R})$
8: **while** $\mathcal{T} \neq \phi$ **do**
9:    $\mathbf{A}_u \leftarrow \mathcal{T}.dequeue()$
10:   **for** each dependent layer $\mathbf{A}_v$ of $\mathbf{A}$ **do**
11:      **if** $\mathbf{A}_v \in \mathcal{L}_u$ **then**
12:         remove edge $\langle \mathbf{A}_u, \mathbf{A}_v \rangle$ from $\mathbf{G}_{Q,0}$
13:         $E_{Q,0} \leftarrow E_{Q,0} \cup \langle \mathbf{A}_u, \mathbf{A}_v \rangle$
14:      **else**
15:         $\mathcal{T}.enqueue(\mathbf{A}_v)$
16:         $\mathcal{L}_v \leftarrow \mathcal{L}_v \cup \mathcal{L}_u \cup \{\mathbf{A}_u\}$
17:      **end if**
18:   **end for**
19: **end while**
20: return $[\mathbf{G}_{Q,0}, E_{Q,0}]$

The algorithms used in Lemma 4.20 together with Lemma 4.19 guarantee that set-ordering preservation property also holds in multi-layered networks with arbitrarily structured dependency graph $\mathbf{G}$.

Now, we are ready to present our main theorem as follows.

Theorem 4.21   Diminishing Returns Property of MuLaN. *For **any** connectivity function $C(\mathbf{A})$ in the* SubLine *family (Eq. (2.1)) and **any** multi-layered network in the* MuLaN *family (Definition (1)); the overall impact of node set $\mathcal{S}_l$ in the control layer l, $\mathbb{I}(\mathcal{S}_l) = \sum_{i=1}^{g} \alpha_i I(\mathcal{S}_{l \to i})$, is (a) monotonically non-decreasing; (b) sub-modular; and (c) normalized.*

Proof. We first prove the sub-modularity of function $\mathbb{I}(\mathcal{S}_l)$. Let $\mathcal{I}_l, \mathcal{J}_l, \mathcal{K}_l$ be three node sets in layer $\mathbf{A}_l$ and $\mathcal{I}_l \subseteq \mathcal{J}_l$. Define the following two sets as: $\mathcal{S}_l = \mathcal{I}_l \cup \mathcal{K}_l$ and $\mathcal{T}_l = \mathcal{J}_l \cup \mathcal{K}_l$. We have that

$$\mathbb{I}(\mathcal{S}_l) - \mathbb{I}(\mathcal{I}_l) = \sum_{i=1}^{g} \alpha_i I(\mathcal{S}_{l\to i}) - \sum_{i=1}^{g} \alpha_i I(\mathcal{I}_{l\to i}) \tag{4.21}$$

$$= \sum_{i=1}^{g} \alpha_i (I(\mathcal{S}_{l\to i}) - I(\mathcal{I}_{l\to i}))$$

$$\mathbb{I}(\mathcal{T}_l) - \mathbb{I}(\mathcal{J}_l) = \sum_{i=1}^{g} \alpha_i I(\mathcal{T}_{l\to i}) - \sum_{i=1}^{g} \alpha_i I(\mathcal{J}_{l\to i}) \tag{4.22}$$

$$= \sum_{i=1}^{g} \alpha_i (I(\mathcal{T}_{l\to i}) - I(\mathcal{J}_{l\to i})).$$

$\forall i = 1, \ldots, g$, it is obvious that $\mathcal{S}_{l\to i} = \mathcal{I}_{l\to i} \cup \mathcal{K}_{l\to i}$, $\mathcal{T}_{l\to i} = \mathcal{J}_{l\to i} \cup \mathcal{K}_{l\to i}$. By Lemma 4.19, we have $\mathcal{I}_{l\to i} \subseteq \mathcal{J}_{l\to i}$. Furthermore, by the sub-modularity of $I(\mathcal{S}_i)$ on $\mathbf{A}_i$ (Lemma 4.18), we have that

$$I(\mathcal{S}_{l\to i}) - I(\mathcal{I}_{l\to i}) \geq I(\mathcal{T}_{l\to i}) - I(\mathcal{J}_{l\to i}).$$

Since for $\forall i$, we have $\alpha_i \geq 0$. Therefore,

$$\sum_{i=1}^{g} \alpha_i (I(\mathcal{S}_{l\to i}) - I(\mathcal{I}_{l\to i})) \geq \sum_{i=1}^{g} \alpha_i (I(\mathcal{T}_{l\to i}) - I(\mathcal{J}_{l\to i})). \tag{4.23}$$

Putting Eqs. (4.21), (4.22), and (4.23) together, we have that

$$\mathbb{I}(\mathcal{S}_l) - \mathbb{I}(\mathcal{I}_l) \geq \mathbb{I}(\mathcal{T}_l) - \mathbb{I}(\mathcal{J}_l)$$

which completes the proof that $\mathbb{I}(\mathcal{S}_l)$ is sub-modular.

Notice that the connectivity function $C(\mathbf{A})$ in the SubLine family is monotonically non decreasing. By Eq. (4.21), we have that

$$\mathbb{I}(\mathcal{S}_l) - \mathbb{I}(\mathcal{I}_l) = \sum_{i=1}^{g} \alpha_i (C(\mathbf{A}_i \setminus \mathcal{I}_l) - C(\mathbf{A}_i \setminus \mathcal{S}_l)) \geq 0$$

which completes the proof that $\mathbb{I}(\mathcal{S}_l)$ is monotonically non-decreasing.

Finally, notice that for each dependent layer, the impact function $I(\mathcal{S}_i)$ is normalize (Lemma 4.18); and for $i = 1, \ldots, g$, $\Phi_{l\to i} = \Phi$ (an empty set). Therefore, we have that $\mathbb{I}(\Phi)$ 0. In other words, $\mathbb{I}(\mathcal{S}_l)$ is also normalized.

4.2.3   PROPOSED ALGORITHM

In this section, we introduce our algorithm to solve OPERA (Problem 4.1), followed by some analysis in terms of the optimization quality as well as the complexity.

A Generic Solution Framework. Finding out the global optimal solution for Problem 4.17 by a brute-force method would be computationally intractable, due to the exponential enumeration. Nonetheless, the diminishing returns property of the impact function $\mathbb{I}(.)$ (Theorem 4.21) immediately lends itself to a greedy algorithm for solving OPERA with any connectivity function in the SUBLINE family and arbitrary member in the MULAN family, as summarized in Algorithm 4.14.

Algorithm 4.14 OPERA: A Generic Solution Framework

Input: (1) A multi-layered network Γ, (2) a control layer $\mathbf{A}_l$, (3) an overall impact function $\mathbb{I}(\mathcal{S}_l)$, and (4) an integer k

Output: A set of k nodes $\mathcal{S}$ from the control layer $\mathbf{A}_l$

  1: initialize $\mathcal{S}$ to be empty
  2: **for** each node v_0 in layer $\mathbf{A}_l$ **do**
  3:    calculate $\text{margin}(v_0) \leftarrow \mathbb{I}(v_0)$
  4: **end for**
  5: find $v = \text{argmax}_{v_0} \text{margin}(v_0)$ and add v to $\mathcal{S}$
  6: set $\text{margin}(v) \leftarrow -1$
  7: **for** $i = 2$ to k **do**
  8:    set $\text{maxMargin} \leftarrow -1$
  9:    **for** each node v_0 in layer $\mathbf{A}_l$ **do**
10:       `/*an optional 'if' for lazy eval.*/`
11:       **if** $\text{margin}(v_0) > \text{maxMargin}$ **then**
12:         calculate $\text{margin}(v_0) \leftarrow \mathbb{I}(\mathcal{S} \cup \{v_0\}) - \mathbb{I}(\mathcal{S})$
13:         **if** $\text{margin}(v_0) > \text{maxMargin}$ **then**
14:           set $\text{maxMargin} \leftarrow \text{margin}(v_0)$ and $v \leftarrow v_0$
15:         **end if**
16:       **end if**
17:    **end for**
18:    add v to $\mathcal{S}$ and set $\text{margin}(v) \leftarrow -1$
19: **end for**
20: return $\mathcal{S}$

In Algorithm 4.14, steps 2–4 calculate the impact score $\mathbb{I}(v_0)$ ($v_0 = 1, 2, ...$) for each node in the control layer $\mathbf{A}_l$. Step 5 selects the node with the maximum impact score. In each iteration in steps 7–19, we select one of the remaining $(k - 1)$ nodes, which would make the maximum marginal increase in terms of the current impact score (step 12, $\text{margin}(v_0) =$

$\mathbb{I}(\mathcal{S} \cup \{v_0\}) - \mathbb{I}(\mathcal{S})$). In order to further speed up the computation, the algorithm admits an *optional* lazy evaluation strategy (adopted from [67]) by activating an optional "if" condition in Step 11.

Note that it is easy to extend Algorithm 4.14 to the scenario where we have multiple control layers. Suppose $\mathcal{A}_l = \{\mathbf{A}_{l_1}, \mathbf{A}_{l_2}, \ldots, \mathbf{A}_{l_x}\}$ is a set of control layers, to select best k nodes from $\mathcal{A}_l$, we only need to scan over all the nodes in $\mathcal{A}_l$ in steps 2 and 9, respectively, and pick the highest impact node from the entire candidate set in steps 5 and 18. Consequently, the resulting set $\mathcal{S}$ returned from the algorithm would contain the k highest impact nodes over $\mathcal{A}_l$.

Algorithm Analysis. Here, we analyze the optimality as well as the complexity of Algorithm 4.14, which are summarized in Lemmas 4.22–4.24. According to these lemmas, our proposed Algorithm 1 leads to a *near-optimal* solution with *linear* complexity.

Lemma 4.22   Near-Optimality. *Let $\mathcal{S}_l$ and $\tilde{\mathcal{S}}_l$ be the sets selected by Algorithm 4.14 and the brute-force algorithm, respectively. Let $\mathbb{I}(\mathcal{S}_l)$ and $\mathbb{I}(\tilde{\mathcal{S}}_l)$ be the overall impact of $\mathcal{S}_l$ and $\tilde{\mathcal{S}}_l$. Then $\mathbb{I}(\mathcal{S}_l) \geq (1 - 1/e)\mathbb{I}(\tilde{\mathcal{S}}_l)$.*

Proof. As proved in Theorem 4.21, the overall impact function $\mathbb{I}(\mathcal{S})$ ($\mathcal{S} \subseteq \mathbf{A}_l$) is monotonic, sub-modular and normalized. Using the property of such functions in [88], we have $\mathbb{I}(\mathcal{S}_l) \geq (1 - 1/e)\mathbb{I}(\tilde{\mathcal{S}}_l)$.   □

Lemma 4.23   Time Complexity. *Let $h(n_i, m_i, |\mathcal{S}_{l \to i}|)$ be the time to compute the impact of node set $\mathcal{S}_l$ on layer i. The time complexity for selecting $\mathcal{S}$ of size k from the control layer $\mathbf{A}_l$ is upper bounded by $O(k(|\mathcal{N}(\mathbf{A}_l)| + n_l \sum_{i=1}^{g} h(n_i, m_i, |\mathcal{S}_{l \to i}|)))$ where $\mathcal{N}(\mathbf{A}_l)$ denotes the nodes and cross-layer links in Γ that depends on $\mathbf{A}_l$.*

Proof. The greedy algorithm with the lazy evaluation strategy needs to iterate over all the nodes in layer $\mathbf{A}_l$ for k time. At each time, the worst case is that we need to evaluate the marginal increase for all unselected nodes in $\mathbf{A}_l$. The overall complexity of finding dependents of every node in $\mathbf{A}_l$ is equal to the size of the sub-system that is on $\mathbf{A}_l$, which is $|\mathcal{N}(\mathbf{A}_l)|$. And for each unselected node, finding out its current impact to the system as shown in steps 3 and 12 can be upper bounded by the complexity of $\sum_{i=1}^{g} h(n_i, m_i, n_i) + g$ because there are at most g non-zero weighted layers that depend on $\mathbf{A}_l$. Taking these all together, the complexity of selecting set $\mathcal{S}$ from $\mathbf{A}_l$ with Algorithm 4.14 is $O(k[|\mathcal{N}(\mathbf{A}_l)| + n_l \sum_{i=1}^{g} h(n_i, m_i, |\mathcal{S}_{l \to i}|)])$, where $|\mathcal{N}(\mathbf{A}_l)|$ is upper bounded by $N + L$, which is the sum of total number of nodes and total number of dependency links in Γ. If given that function h is linear to n_i, m_i and $|\mathcal{S}_{l \to i}|$, as $|\mathcal{S}_{l \to i}|$ is upper bounded by n_i, and n_l can be viewed as a constant compared to N, M, and L, it is easy to see that the complexity of the algorithm is linear to N, M, and L.

Remarks. Lemma 4.23 implies a linear time complexity of the proposed OPERA algorithm w.r.t. the size of the entire multi-layered network ($N + M + L$), where N, M, L are the total number of nodes, the total number of within-layer links and the total number of cross-layer links in Γ under the condition that the function h is linear w.r.t. n_i, m_i, and $|\mathcal{S}_{l\to i}|$. This condition holds for most of the network connectivity measures in the SUBLINE family, e.g., the path capacity, the truncated loop capacity, and the triangle capacity. To see this, let us take the most expensive truncated loop capacity as an example. The time complexity for calculating truncated loop capacity in a single network is $O(mr + nr^2)$, where r is the number of eigenvalues used in the calculation and it is often orders of magnitude smaller compared with m and n. On the other hand, we have $|\mathcal{N}(\mathbf{A}_l)| \leq N + L$. Therefore, the overall time complexity for selecting set $\mathcal{S}$ of size k from control layer $\mathbf{A}_l$ is upper bounded by $O(k(N + L + n_l \sum_{i=1}^{g}(m_i r + n_i r^2))) = O(k(N + L + n_l(rM + r^2 N))) = O(k(L + n_l(rM + r^2 N)))$.

Lemma 4.24   Space Complexity. *Let $w(n_i, m_i, |\mathcal{S}_{l\to i}|)$ be a function of n_i, m_i and $|\mathcal{S}_{l\to i}|$ that denotes the space cost to compute $I(\mathcal{S}_{l\to i})$. The space complexity of Algorithm 4.14 is $O(N + M + L)$ under the condition that the function w is linear w.r.t. n_i, m_i, and $|\mathcal{S}_{l\to i}|$.*

Proof. As defined in Lemma 4.23, N, M, and L are the total number of nodes, total number of within-layer links and total number of cross-layer links in Γ. Then storing multi-layered network Γ would take a space of $O(N + M + L)$. In Algorithm 4.14, it takes $O(n_l)$ to save the marginal increase vector (*margin*) and $O(k)$ to save result $\mathcal{S}$. As space for computing $I(\mathcal{S}_{l\to i})$ can be reused for each layer i, then computing $\mathbb{I}(\mathcal{S}_{l\to i})$ is bounded by $\arg max_i\, w(n_i, m_i, |\mathcal{S}_{l\to i}|)$. If function w is linear w.r.t. n_i, m_i and $|\mathcal{S}_{l\to i}|$, then the space complexity of Algorithm 4.14 is of $O(N + M + L + k + n_l) + O(\arg max_i(n_i)) + O(\arg max_i(m_i)) = O(N + M + L)$. □

Remarks. The condition that the function w is linear w.r.t. n_i, m_i, and $|\mathcal{S}_{l\to i}|$ holds for most of the network connectivity measures in the SUBLINE family, which in turn implies a linear space complexity for the proposed OPERA algorithm. Again, let us take the truncated loop capacity connectivity as an example. Storing the input MULAN (Γ) takes $O(N + M + L)$ in space. The space cost to calculate the truncated loop capacity in a single-layered network is $O(m + nr)$, where r is the number of eigenvalues used for the computation. Again, r is usually a much smaller number compared with m and n, and thus is considered as a constant. Therefore, the overall space complexity for OPERA with the truncated loop capacity is $O(N + M + L)$.

4.2.4   EXPERIMENTAL EVALUATION

In this section, we empirically evaluate the proposed OPERA algorithms. All experiments are designed to answer the following two questions:

Table 4.6: Data sets summary

Data Sets	Application Domains	Number of Layers	Number of Nodes	Number of Links
D1	MultiAS	2~4	5,929~24,539	11,183~50,778
D2	InfraNet	3	19,235	46,926
D3	SocInNet	2	63,501~124,445	13,097~211,776
D4	BIO	3	35,631	253,827

- *Effectiveness*: how effective are the proposed OPERA algorithms at optimizing the connectivity measures (defined in the proposed SUBLINE family) of a multi-layered network (from the proposed MULAN family)?

- *Efficiency*: how fast and scalable are our algorithms?

Experimental Setup

Data Sets Summary. We perform the evaluations on four different application domains, including (D1) a multi-layered Internet topology at the autonomous system level (MULTIAS); (D2) critical infrastructure networks (INFRANET); (D3) a social-information collaboration network (SOCINNET); and (D4) a biological CTD (Comparative Toxicogenomics Database) network [30] (BIO). For the first two domains, we use real networks to construct the within-layer networks (i.e., $\mathcal{A}$ in the MULAN model) and construct one or more cross-layer dependency structures based on real application scenarios (i.e., **G** and $\mathcal{D}$ in the MULAN model). For the data sets in SOCINNET and BIO domains, both the within-layer networks and cross-layer dependency networks are based on real connections. A summary of these data sets is shown in Table 4.6. We will present the detailed description of each application domains in Section 4.2.4

Baseline Methods. To our best knowledge, there is no existing method that can be directly applied to the connectivity optimization problem (Problem 4.1) of the MULAN model. We generate the baseline methods using two complementary strategies, including *forward propagation* ("FP" for short) and *backward propagation* ("BP" for short). The key idea behind the *forward propagation* strategy is that an important node in the *control layer* might have more impact on its dependent networks as well. On the other hand, for the *backward propagation strategy*, we first identify important nodes in the *target layer(s)*, and then trace back to its supporting layer(s) through the cross-layer dependency links (i.e., $\mathcal{D}$). For both strategies, we need a node importance measure. In our evaluations, we compare three different measures, including (1) node degree; (2) PageRank measure [94]; and (3) *Netshield* values [124]. In addition, for comparison purposes, we also randomly select nodes either from the control layer (for the forward propagation strategy) or from the target layer(s) (for the backward propagation strategy). Altogether, we have eight baseline methods (four for each strategy, respectively), including (1) "Degree-FP," (2) "PageRank-FP," (3) "Netshield-FP," (4) "Rand-FP," (5) "Degree-BP," (6) "PageRank-BP," (7) "Netshield-BP," and (8) "Rand-BP."

Opera Algorithms and Variants. We evaluate three prevalent network connectivity measures, including (1) the leading eigenvalue of the (within-layer) adjacency matrix, which relates to the epidemic threshold of a variety of cascading models; (2) the loop capacity (LC), which relates to the robustness of the network; and (3) the triangle capacity (TC), which relates to the local connectivity of the network. As mentioned in Chapter 2, both the loop capacity and the triangle capacity are members of the SUBLINE family. Strictly speaking, the leading eigenvalue does *not* belong to the SUBLINE family. Instead, it approximates the path capacity (PC), and the latter (PC) is a member of the SUBLINE family. Correspondingly, we have three instances of the proposed OPERA algorithm (each corresponding to one specific connectivity measure) including "OPERA-PC," "OPERA-LC," and "OPERA-TC." Recall that there is an optional lazy evaluation step (step 11) in the proposed OPERA algorithm, thanks to the diminishing returns property of the SUBLINE connectivity measures. When the leading eigenvalue is chosen as the connectivity function, such diminishing returns property does not hold anymore. To address this issue, we introduce a variant of OPERA-PC as follows. At each iteration, after the algorithm chooses a new node v (step 18, Algorithm 4.14), we (1) update the network by removing all the nodes that depend on node v, and (2) update the corresponding leading eigenvalues and eigenvectors. We refer to this variant as "OPERA-PC-Up." For each of the three connectivity measures, we run all four OPERA algorithms.

Machines and Repeatability. All the experiments are performed on a machine with two processors Intel Xeon 3.5 GHz with 256 GB of RAM. The algorithms are programmed with MATLAB using a single thread. All the data sets used in this work are publicly available.

Effectiveness Results

D1 – MultiAS. This data set contains the Internet topology at the autonomous system level. The data set is available at http://snap.stanford.edu/data/. It has 9 different network snapshots, with $633 \sim 13{,}947$ nodes and $1{,}086 \sim 30{,}584$ edges. In our evaluations, we treat these snapshots as the within-layer adjacency matrices $\mathcal{A}$. For a given supporting layer, we generate the cross-layer node-node dependency matrices $\mathcal{D}$ by randomly choosing three nodes from its dependent layer as the direct dependents for each supporting node. For this application domain, we have experimented with different layer-layer dependency structures (**G**), including a three-layered line-structured network, a three-layered tree-structured network, a four-layered diamond-shaped network, and a three-layered cyclic network. As the experimental results in the first three networks follow a similar pattern, we only present the results on the diamond-shaped network and cyclic network in Figures 4.16 and 4.17 due to page limits. Overall, the four instances of the proposed OPERA algorithm perform better than the baseline methods. Among the baseline

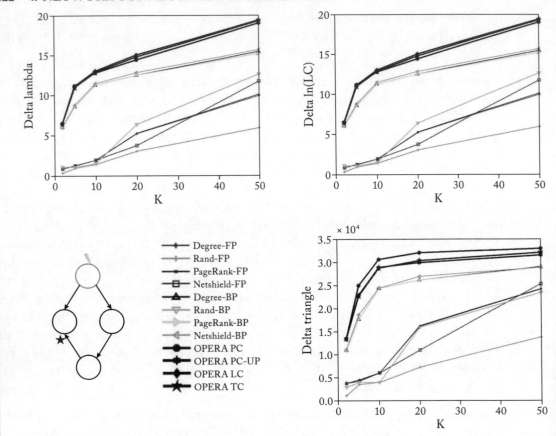

Figure 4.16: Evaluations on the MULTIAS data set, with a four-layered diamond-shaped dependency network. The connectivity change vs. budget. Larger is better. All four instances of the proposed OPERA algorithm (in red) outperform the baseline methods.

methods, the *backward propagation* methods are better than the forward propagation methods under acyclic dependency networks (Figure 4.16). This is because the length of the backtracking path on the dependency network **G** (from the target layer to the control layer) is short. Therefore, compared with other baseline methods, the node set returned from the BP strategy is able to affect more important nodes in the target layer. While for the cyclic dependency networks in Figure 4.17, the backtracking path is elongated by the cycle. Then the nodes selected by BP strategy are not guaranteed to affect more important nodes in the target layer than FP strategy.

D2 – InfraNet. This data set contains three types of critical infrastructure networks, including (1) the power grid, (2) the communication network, and (3) the airport networks. The power grid is an undirected, unweighted network representing the topology of the Western States Power

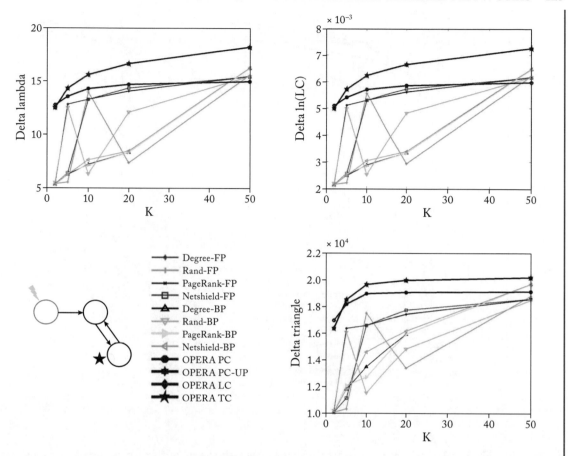

Figure 4.17: Evaluations on the MULTIAS data set, with a three-layered cyclic dependency network. The connectivity change vs. budget. Larger is better. Three out of four instances of the proposed OPERA algorithm (in red) outperform the baseline methods.

Grid of the United State [133]. It has 4,941 nodes and 6,594 edges. We use one snapshot from the MULTIAS data set as the communication network with 11,461 nodes and 32,730 edges. The airport network represents the internal U.S. air traffic lines between 2,649 airports and has 13,106 links (available at http://www.levmuchnik.net/Content/Networks/NetworkData.html). We construct a triangle-shaped layer-layer dependency network **G** (see the icon of Figure 4.18) based on the following observation. The operation of an airport depends on both the electricity provided by the power grid and the Internet support provided by the communication network. In the meanwhile, the full-functioning of the communication network depends on the support of the power grid. We use a similar strategy as in MULTIAS to generate the cross-layer node-node dependency matrices $\mathcal{D}$. The results are summarized in Figure 4.18. Again, the proposed

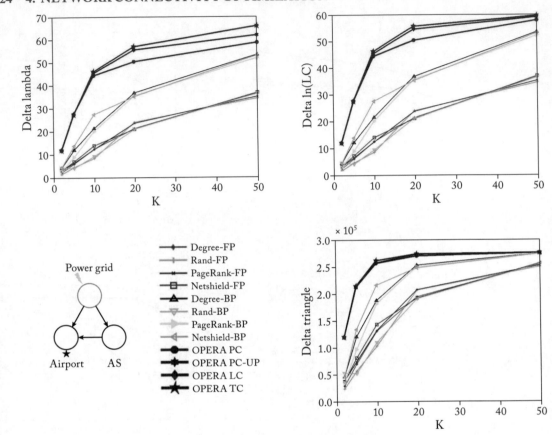

Figure 4.18: Evaluations on the INFRANET data set, with a three-layered triangle-shaped dependency network. The connectivity change vs. budget. Larger is better. All the four instances of the proposed OPERA algorithm (in red) outperform the baseline methods.

OPERA algorithms outperform all the baseline methods. Similar to the MULTIAS network, the backtracking path from the airport layer to the power grid layer is also very short. Therefore, the backward propagation strategies perform relatively better than other baseline methods. In addition, we change the density of the cross-layer node-node dependency matrices and evaluate its impact on the optimization results in Figure 4.19. We found that (1) across different dependency densities, the proposed OPERA algorithms still outperform the baseline methods and (2) when the dependency density increases, the algorithms lead to a larger decrease of the corresponding connectivity measures with the same budget.

D3 – SocInNet. This data set contains three types of social-information networks [117], including (1) a co-authorship network; (2) a paper-paper citation network; and (3) a venue-venue

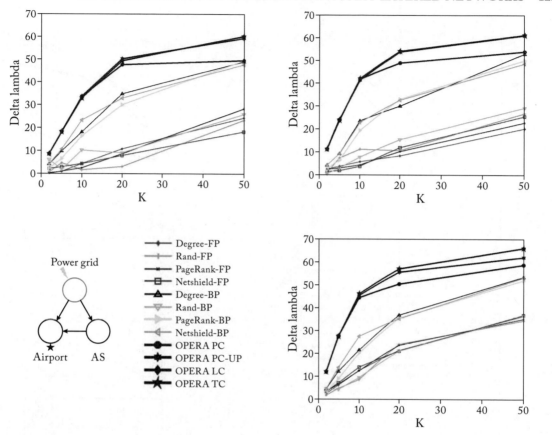

Figure 4.19: $\Delta\lambda$ w.r.t. k. Change the average number of dependents between power grid and as from 5, 10, to 15 (left to right).

citation network. Different from the previous two data sets, two types of cross-layer node-node dependency links naturally exist in this data set, including *who-writes-which paper*, and *which venue-publishes-which paper*. In our experiment, we use the papers published between year 1990–1992. In total, there are 62,602 papers, 61,843 authors, 899 venues, 10,739 citation links, 201,037 collaboration links, 2,358 venue citation links, 126,242 author-paper cross-layer links, and 62,602 venue-paper cross-layer links.

We evaluate the proposed algorithms in two scenarios with this data set, including (1) an author-paper two-layered network and (2) a venue-paper two-layered network. For both scenarios, we choose the paper-paper citation network as the target layer. Figure 4.20 presents the results on the author-paper two-layered network. We can see that three out of four OPERA algorithms outperform all the baseline methods in all three cases. OPERA-PC does not perform

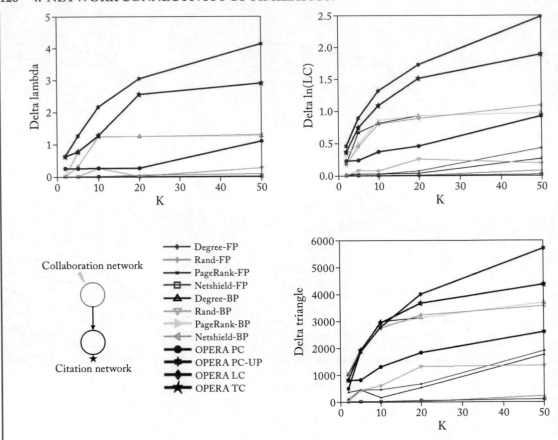

Figure 4.20: Evaluations on the SocInNet data set, with a two-layered author-paper dependency network. The connectivity change vs. budget. Larger is better. Three out of four proposed Opera algorithms (in red) outperform the baseline methods.

as well as the remaining three Opera instances due to the gap between the leading eigenvalu and the actual path capacity. However, the issue can be partially addressed with Opera-PC-U by introducing an update step. Among the baseline methods, the *backward propagation* strat egy is better since the target layer is directly dependent on the control layer, which makes possible to trace back the high-impact authors given the set of high-impact papers. The po performance of the *forward propagation* methods implies that a socially active author does n necessarily have high-impact papers. The results on the venue-paper network is similar as show in Figure 4.21. Different from the author-paper network, the backward propagation strategi perform worse than the forward propagation strategies. This is probably due to the fact that n all the important (high-impact) papers appear in the important (high-impact) venues.

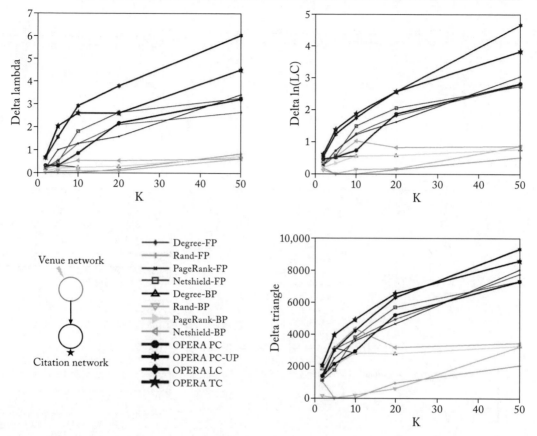

Figure 4.21: Evaluations on the SocInNet data set, with a two-layered venue-paper dependency network. The connectivity change vs. budget. Larger is better. Three out of four proposed Opera algorithms (in red) outperform the baseline methods.

D4 – BIO. This data set contains three types of biological networks [30] including (1) a chemical similarity network with 6,026 chemicals, 69,109 links; (2) a gene similarity network with 25,394 genes, 154,167 links; and (3) a disease similarity network with 4,256 diseases, 30,551 links. The dependencies between those layers depict *which chemical-affects-which gene, which chemical-treats-which disease*, and *which gene-associates-which disease* relations, each of which contains 53,735, 19,771, and 1,950 dependency links, respectively. The evaluation results are as shown in Figure 4.22. Despite the fact that the proposed Opera algorithms outperform all other baseline methods, there are two interesting observations that are worth to be mentioned. First is that the impact of chemical nodes on disease networks becomes saturated at a small budge value for all connectivity measures, which implies that only a few chemicals are effective in treating

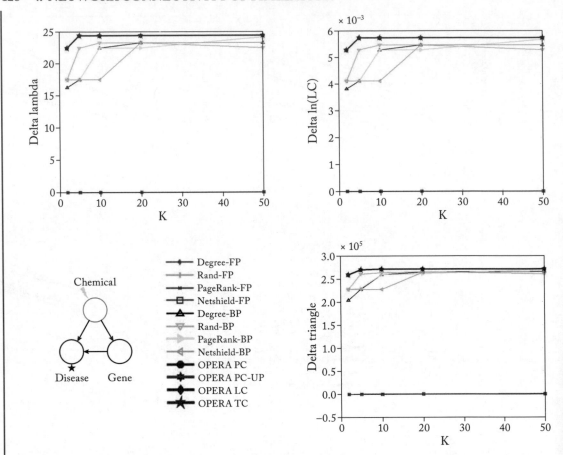

Figure 4.22: Evaluations on the BIO data set, with a three-layered triangle-shaped dependency network. The connectivity change vs. budget. Larger is better. All four proposed OPERA algorithms (in red) outperform the baseline methods.

most of the diseases in the given data set. Second, the ineffectiveness of *forward propagation* methods indicates that chemicals with various compounds (high within-layer centrality nodes) may have little effect in disease treatment.

Efficiency Results

Figure 4.23 presents the scalability of the proposed OPERA algorithms. We can see that all four instances of OPERA scale linearly with respect to the size of the input multi-layered network (i.e., $N + M + L$), which is consistent with our complexity analysis. The wall-clock time for OPERA

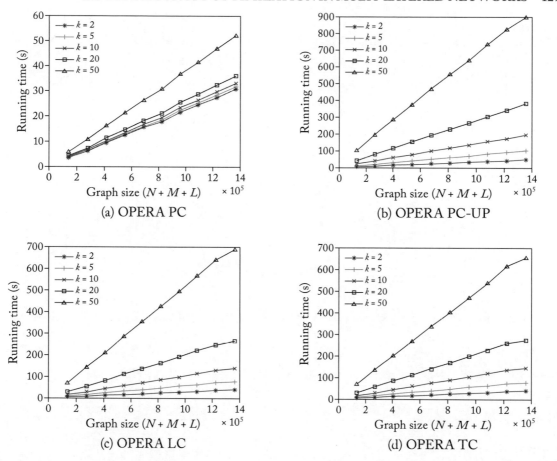

Figure 4.23: Wall-clock time vs. The size of the input networks. The proposed OPERA algorithms scale linearly w.r.t. $(N + M + L)$.

PC-Up is the longest compared with the remaining three instances, due to the additional update step.

CHAPTER 5

Conclusion and Future Work

In this chapter, we summarize our key research results and discuss future research directions for network connectivity.

5.1 CONCLUSION

In this book, we propose three main tasks for network connectivity studies: (1) connectivity measure concepts, (2) connectivity inference computation, and (3) connectivity optimization.

Measure Concepts. For connectivity measures, our main finding is that various task-oriented connectivity measures in the literature can be unified into a generalized model—the SUBLINE connectivity model. The key idea of SUBLINE model is to view the connectivity of the network as the aggregation of the connectivity of some valid subgraphs. By restricting the valid subgraphs to a subset of nodes, the SUBLINE connectivity can be used to measure the local connectivity of a subnetwork. Moreover, we also show that the proposed model can be easily extended to multi-layered networks.

Inference Computation. For connectivity inference, we addressed the eigen-functions/connectivity tracking problem in dynamic networks and the cross-layer dependency inference problem in multi-layered networks. To efficiently track the eigen-functions in the network, we propose TRIP-BASIC and TRIP. In addition, we provide a framework for attribution analysis on eigen-functions and a method to effectively estimate tracking errors. Our experiments show that both Trip-Basic and Trip can effectively and efficiently track the changes of eigen-pairs, the number of triangles, robustness score, and eigen-gap in dynamic graphs, while Trip is more stable over time. In both cases, the accumulated error rate inevitably keeps increasing as time goes by. As for the dependency inference problem, we propose to formulate the inference problem as a collective collaborative filtering problem and introduce FASCINATE, an algorithm that can effectively infer the missing dependencies with provable optimality and scalability. In particular, by modeling the impact of the zero-start node as a perturbation in the multi-layered network, we derive FASCINATE-ZERO, an online variant of FASCINATE that can approximate the dependencies of the newly added node with sub-linear complexity w.r.t. the overall system size. The experimental results on five real-world datasets demonstrate the superiority of our proposed algorithm both by its effectiveness and efficiency.

Connectivity Optimization. For the connectivity optimization task, we first prove that for any network connectivity measures in the SubLine family, the connectivity optimization problem with the MuLaN model enjoys the diminishing returns property, which naturally lends itself to a family of provable near-optimal algorithms using the greedy scheme. Then we show that a wide range of network connectivity optimization (NETCOM) problems are NP-complete and $(1 - 1/e)$ is the best approximation ratio that a polynomial algorithm can achieve for NETCOM problems unless $NP \subseteq DTIME(n^{O(\log\log n)})$. On the algorithmic aspect, we propose a series of effective, scalable, and generalizable optimization algorithms CONTAIN and Opera that can be applied to both single-layered networks and multi-layered networks.

5.2   FUTURE RESEARCH DIRECTIONS

Network connectivity is a powerful graph parameter that may lead to many interesting findings. Below we present some promising research directions.

5.2.1   COMPLEX MULTI-LAYERED NETWORK CONNECTIVITY

In our work, we assume that the node in a multi-layered network is able to function as long as all of its cross-layer support nodes are functioning and the node itself is connected to its within-layer network. Under this assumption, we model the connectivity of a multi-layered network as an aggregation over the connectivity on each layer. However, the node functioning constraints can be far more complicated in some applications. In certain power grid networks, the functioning of a power station would require the node to be connected to the *largest connected component* in its within-layer network [12, 41, 48, 96]. One simple example of such constraint is shown in Figure 5.1. Here, the power stations p_1 and p_3 cannot be considered as functioning nodes under this constraint even though they are connected to each other, as they are not connected to the largest connected component (circled in black) in the power grid. On the other hand, the cross-layer dependency relationships may impose sophisticated constraints to node functioning as well. For example, in Figure 5.1, the functioning of node v may require the functioning of *all* of its supporting nodes p_1, p_2, p_3 (the dependency type used in this book), or the functioning of *any* of its supporting nodes. The two different dependency relationships would significantly affect the robustness of the resulting systems.

In the existing literature, extensive efforts have been made to study the multi-layered network connectivity under different within-layer and cross-layer functioning constraints. In [91] Nguyen et al. studied the connectivity in a two-layered network model where the functioning of the node is determined by (1) its connection to the largest connected component of its within-layer network, and (2) the functioning of *any* of its support nodes. While in [3, 108], the connectivity in a new two-layered network model was studied, where the functioning of each node is determined by the functioning of its support nodes through a customized Boolean logic function.

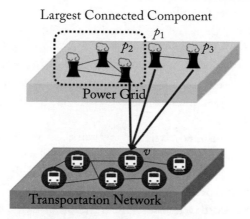

Figure 5.1: An illustrative example of complex multi-layered network functioning constraints.

Clearly, the diverse node functioning constraints have led to different definitions of connectivity in multi-layered networks. Consequently, the corresponding connectivity optimization strategies would vary from one to another significantly. Moreover, the existing works are predominantly based on two-layered networks. Thus, it would be an interesting problem to investigate the commonality among the multi-layered network connectivity models and the generalization to arbitrarily interdependent multi-layered networks.

5.2.2 DYNAMIC NETWORK INFERENCE

Real-world networks are evolving over time. In the infrastructure system, new power stations are constantly being added to the system to fulfill the increasing need for electricity supply. Similar expanding processes can be observed in autonomous systems and transportation networks as well. Simultaneously, new cross-layer dependencies across those layers would be established in the system. The changing structure would inevitably affect the inference results from the old system. Therefore, it is necessary to re-calibrate the inference results timely to accommodate system changes. The naive way to solve the dynamic network inference problem is to re-run the static inference algorithms whenever the system changes. However, such a strategy would be very inefficient when the system is changing fast.

Recent advances for the dynamic network inference problem have been summarized in several survey papers [1, 52]. Particularly, our incremental decomposition algorithm, as the core part of our eigen-functions tracking framework in Chapter 3, has been successfully extended to address the dynamic network link prediction problem [130] and the dynamic attributed network embedding problem [70]. In [68], Levin et al. proposed a method that can efficiently re-calibrate the adjacency spectral embedding when a new node is added to the network. To control the accumulated approximation error in such incremental update dynamic systems, Zhang et al. [143]

proposed an error tracking algorithm, which can notify at the time stamp where the accumulated error is greater than the predefined threshold. In [21], the incremental decompositio tracking scheme is also extended to address the dynamic multi-layered network inference prob lem. Other than the above incremental decomposition-based network tracking works, man dynamic node representation learning algorithms are proposed based on random walk base models ([8, 80, 105]), sequential models ([81, 84, 87, 109, 142]), auto-encoder based mode ([11, 42, 43, 99]), etc.

The above-mentioned dynamic network inference works are mostly based on single layered network models. Their extensions to multi-layered networks are often non-trivial du to the diverse node types and edge semantics. Specifically, the nodes from different layers ma come from different types, and the evolution of cross-layer dependencies and within-layer lin may be attributed to different causes. On the other hand, the cross-layer dependency chang and within-layer link changes may be correlated intrinsically as well. Thus, how to effective model the dynamic of multi-layered networks would be a critical question to explore for th dynamic network inference problem.

5.2.3 CONNECTIVITY OPTIMIZATION AND ADVERSARIAL ATTACK

The adversarial attack on networked data has become a trending topic in recent years. Its ma idea is to alter the network structure and node attributes to affect the results of subsequent tas to the maximum extent.

In [29], Dai et al. studied the attack on node classification and graph classification tas by manipulating the edges in the network. In [145], Zügner et al. proposed an attack strate; that manipulates both edges and node features in the graph subject to certain constraint conc tions, so that the perturbations are unnoticeable to the original network. Aside from changi edges and node features as attack operations, injecting vicious nodes to the network is also i vestigated in [116, 129] to mislead the graph learning models. In [10], the poisoning attack unsupervised node embedding was studied by leveraging matrix perturbation theory. A mo comprehensive literature review can be found in [115] and [50].

One interesting observation for the adversarial attack on networked data and netwo connectivity optimization is that both problems are trying to manipulate the network structu to affect the network connectivity/graph learning results. Thus, one natural question we want ask here would be whether the connectivity of the network is correlated with the robustness graph learning results. Correspondingly, it would be intriguing to know whether the nodes/edg selected by the network connectivity optimization algorithms are critical to adversarial atta tasks as well.

5.2.4 CONNECTIVITY ON HIGH-ORDER DEPENDENCY NETWORKS

Networks are extensively used to model the direct connection between nodes. When generati a walk from the network (i.e., a sequence of nodes to visit along the network edges), m

(a) First-order dependency network   (b) High-order dependency network

Figure 5.2: An illustrative example of shipping traffic network from the first-order dependency perspective and the high-order dependency perspective. Adapted from [137].

algorithms implicitly assume that the Markov property holds in the process. In other words, standing at the current node v in the network, the next step is only determined by the direct (i.e., first-order) connections of node v, rather than the prior steps that lead to node v. The first-order Markov assumption can be very limited for sequential data analysis in flow networks, such as the user behavior analysis in web clickstream networks and traffic study in transportation networks [61, 103, 137]. Consider the shipping traffic network in Figure 5.2 as an example, where the traffic volume between cities is represented by the edge weight in the network (i.e., arrow width in the Figure). The first-order dependency network in Figure 5.2a indicates that a ship departing from the intermediate city *Singapore* may end up in *Los Angeles* or *Seattle* of equal probabilities. However, by the high-order dependency network representation in Figure 5.2b, it is clear to see that ships originating from *Tokyo* are more likely to choose *Los Angeles* as their final destination.

Various algorithms have been proposed to extract the high-order dependency paths in the network and were successfully applied to anomaly detection and temporal network prediction tasks [104, 107, 137]. However, the connectivity problems on such high-order dependency networks still remain largely open. Particularly, the general path capacity, triangle capacity, or loop capacity-based connectivity measures defined in Chapter 2 may no longer be good fits to the high-order dependency scenario, as the extracted dependency paths may vary by applications. To properly define the connectivity in such networks, we can resort to the SuBLINE connectivity measures in Chapter 2 as in Eq. (5.1):

$$C(G, f) = \sum_{\pi \subseteq G} f(\pi). \tag{5.1}$$

Specifically, the valid subgraphs (i.e., π) correspond to all the extracted high-order dependency paths in the network; while $f(\pi)$ may map the dependency path π to an importance score (e.g., path dependence confidence score or total flow volume along the path). The resulting connectivity measure can be used as a metric to identify critical nodes or edges in high-order dependency networks.

Bibliography

[1] Charu Aggarwal and Karthik Subbian. Evolutionary network analysis: A survey. *ACM Computing Surveys (CSUR)*, 47(1):1–36, 2014. DOI: 10.1145/2601412

[2] Réka Albert, Hawoong Jeong, and Albert-László Barabási. Error and attack tolerance of complex networks. *Nature*, 406(6794):378–382, 2000. DOI: 10.1038/35019019

[3] Joydeep Banerjee, Chenyang Zhou, Arun Das, and Arunabha Sen. On robustness in multilayer interdependent networks. *International Conference on Critical Information Infrastructures Security*, pages 247–250, Springer, 2015. DOI: 10.1007/978-3-319-33331-1_21

[4] Albert-László Barabási and Réka Albert. Emergence of scaling in random networks. *Science*, 286(5439):509–512, 1999. DOI: 10.1126/science.286.5439.509

[5] Federico Battiston, Vincenzo Nicosia, and Vito Latora. Structural measures for multiplex networks. *Physical Review E*, 89(3):032804, 2014. DOI: 10.1103/physreve.89.032804

[6] Austin R. Benson, Rediet Abebe, Michael T. Schaub, Ali Jadbabaie, and Jon Kleinberg. Simplicial closure and higher-order link prediction. *ArXiv Preprint ArXiv:1802.06916*, 2018. DOI: 10.1073/pnas.1800683115

[7] Michele Berlingerio, Michele Coscia, Fosca Giannotti, Anna Monreale, and Dino Pedreschi. Foundations of multidimensional network analysis. *Advances in Social Networks Analysis and Mining (ASONAM), International Conference on*, pages 485–489, IEEE, 2011. DOI: 10.1109/asonam.2011.103

[8] Ranran Bian, Yun Sing Koh, Gillian Dobbie, and Anna Divoli. Network embedding and change modeling in dynamic heterogeneous networks. *Proc. of the 42nd International ACM SIGIR Conference on Research and Development in Information Retrieval*, pages 861–864, 2019. DOI: 10.1145/3331184.3331273

[9] Stefano Boccaletti, Ginestra Bianconi, Regino Criado, Charo I. Del Genio, Jesús Gómez-Gardenes, Miguel Romance, Irene Sendina-Nadal, Zhen Wang, and Massimiliano Zanin. The structure and dynamics of multilayer networks. *Physics Reports*, 544(1):1–122, 2014. DOI: 10.1016/j.physrep.2014.07.001

[10] Aleksandar Bojchevski and Stephan Günnemann. Adversarial attacks on node embeddings via graph poisoning. *International Conference on Machine Learning*, pages 695–704, PMLR, 2019.

[11] Stephen Bonner, John Brennan, Ibad Kureshi, Georgios Theodoropoulos, Andrew Stephen McGough, and Boguslaw Obara. Temporal graph offset reconstruction: Towards temporally robust graph representation learning. *IEEE International Conference on Big Data (Big Data)*, pages 3737–3746, 2018. DOI: 10.1109/bigdata.2018.8622636

[12] Sergey V. Buldyrev, Roni Parshani, Gerald Paul, H. Eugene Stanley, and Shlomo Havlin. Catastrophic cascade of failures in interdependent networks. *Nature*, 464(7291):1025–1028, 2010. DOI: 10.1038/nature08932

[13] Deepayan Chakrabarti, Yang Wang, Chenxi Wang, Jurij Leskovec, and Christos Faloutsos. Epidemic thresholds in real networks. *ACM Transactions on Information and System Security (TISSEC)*, 10(4):1, 2008. DOI: 10.1145/1284680.1284681

[14] Hau Chan, Leman Akoglu, and Hanghang Tong. Make it or break it: Manipulating robustness in large networks. *Proc. of SIAM International Conference on Data Mining*, pages 325–333, 2014. DOI: 10.1137/1.9781611973440.37

[15] Chen Chen, Jingrui He, Nadya Bliss, and Hanghang Tong. On the connectivity of multi-layered networks: Models, measures, and optimal control. *Data Mining (ICDM), IEEE 15th International Conference on*, pages 715–720, 2015. DOI: 10.1109/icdm.2015.104

[16] Chen Chen, Ruiyue Peng, Lei Ying, and Hanghang Tong. Network connectivity optimization: Fundamental limits and effective algorithms. *Proc. of the 24th ACM SIGKDD International Conference on Knowledge Discovery and Data Mining*, pages 1167–1176 2018. DOI: 10.1145/3219819.3220019

[17] Chen Chen and Hanghang Tong. On the eigen-functions of dynamic graphs: Fast tracking and attribution algorithms. *Statistical Analysis and Data Mining: The ASA Data Science Journal*, 10(2):121–135, 2017. DOI: 10.1002/sam.11310

[18] Chen Chen, Hanghang Tong, B. Aditya Prakash, Tina Eliassi-Rad, Michalis Faloutsos, and Christos Faloutsos. Eigen-optimization on large graphs by edge manipulation *TKDD*, 10(4):49, 2016. DOI: 10.1145/2903148

[19] Chen Chen, Hanghang Tong, B. Aditya Prakash, Charalampos E. Tsourakakis, Tina Eliassi-Rad, Christos Faloutsos, and Duen Horng Chau. Node immunization on large graphs: Theory and algorithms. *IEEE Transactions on Knowledge and Data Engineering* 28(1):113–126, 2016. DOI: 10.1109/tkde.2015.2465378

[20] Chen Chen, Hanghang Tong, Lei Xie, Lei Ying, and Qing He. FASCINATE: Fast cross-layer dependency inference on multi-layered networks. *Proc. of the 22nd ACM SIGKDD International Conference on Knowledge Discovery and Data Mining*, pages 765–774, San Francisco, CA, August 13–17, 2016. DOI: 10.1145/2939672.2939784

[21] Chen Chen, Yinglong Xia, Hui Zang, Jundong Li, Huan Liu, and Hanghang Tong. Incremental one-class collaborative filtering with co-evolving side networks. *Knowledge and Information Systems*, 63(1):105–124, 2021. DOI: 10.1007/s10115-020-01511-x

[22] Liangzhe Chen, Xinfeng Xu, Sangkeun Lee, Sisi Duan, Alfonso G. Tarditi, Supriya Chinthavali, and B. Aditya Prakash. Hotspots: Failure cascades on heterogeneous critical infrastructure networks. *Proc. of the ACM on Conference on Information and Knowledge Management*, pages 1599–1607, 2017. DOI: 10.1145/3132847.3132867

[23] Pin-Yu Chen and Alfred O. Hero. Local fiedler vector centrality for detection of deep and overlapping communities in networks. *Acoustics, Speech and Signal Processing (ICASSP), IEEE International Conference on*, pages 1120–1124, 2014. DOI: 10.1109/icassp.2014.6853771

[24] Wei Chen, Wynne Hsu, and Mong Li Lee. Making recommendations from multiple domains. *Proc. of the 19th ACM SIGKDD International Conference on Knowledge Discovery and Data Mining*, pages 892–900, 2013. DOI: 10.1145/2487575.2487638

[25] Wei Chen, Chi Wang, and Yajun Wang. Scalable influence maximization for prevalent viral marketing in large-scale social networks. *Proc. of the 16th ACM SIGKDD International Conference on Knowledge Discovery and Data Mining*, pages 1029–1038, 2010. DOI: 10.1145/1835804.1835934

[26] Wei Chen, Yajun Wang, and Siyu Yang. Efficient influence maximization in social networks. *Proc. of the 15th ACM SIGKDD International Conference on Knowledge Discovery and Data Mining*, pages 199–208, 2009. DOI: 10.1145/1557019.1557047

[27] Fan R. K. Chung. *Spectral Graph Theory*, vol. 92, American Mathematical Society, 1997. DOI: 10.1090/cbms/092

[28] Reuven Cohen, Shlomo Havlin, and Daniel Ben-Avraham. Efficient immunization strategies for computer networks and populations. *Physical Review Letters*, 91(24):247901, 2003. DOI: 10.1103/physrevlett.91.247901

[29] Hanjun Dai, Hui Li, Tian Tian, Xin Huang, Lin Wang, Jun Zhu, and Le Song. Adversarial attack on graph structured data. *International Conference on Machine Learning*, pages 1115–1124, PMLR, 2018.

140 BIBLIOGRAPHY

[30] Allan Peter Davis, Cynthia J. Grondin, Kelley Lennon-Hopkins, Cynthia Saraceni-Richards, Daniela Sciaky, Benjamin L. King, Thomas C. Wiegers, and Carolyn J. Mattingly. The comparative toxicogenomics database's 10th year anniversary: update 2015. *Nucleic Acids Research*, 43(D1):D914–D920, 2015.

[31] Manlio De Domenico, Albert Solé-Ribalta, Emanuele Cozzo, Mikko Kivelä, Yamir Moreno, Mason A. Porter, Sergio Gómez, and Alex Arenas. Mathematical formulation of multilayer networks. *Physical Review X*, 3(4):041022, 2013. DOI: 10.1103/physrevx.3.041022

[32] Manlio De Domenico, Albert Solé-Ribalta, Elisa Omodei, Sergio Gómez, and Alex Arenas. Ranking in interconnected multilayer networks reveals versatile nodes. *Nature Communications*, 6, 2015. DOI: 10.1038/ncomms7868

[33] Chris Ding, Tao Li, Wei Peng, and Haesun Park. Orthogonal nonnegative matrix t-factorizations for clustering. *Proc. of the 12th ACM SIGKDD International Conference on Knowledge Discovery and Data Mining*, pages 126–135, 2006. DOI: 10.1145/1150402.1150420

[34] Petros Drineas and Michael W. Mahoney. On the Nyström method for approximating a gram matrix for improved kernel-based learning. *The Journal of Machine Learning Research*, 6:2153–2175, 2005.

[35] Paul Erdos, Alfréd Rényi, et al. On the evolution of random graphs. *Publications of the Mathematical Institute of the Hungarian Academy of Sciences*, 5(1):17–60, 1960. DOI: 10.1515/9781400841356.38

[36] Michalis Faloutsos, Petros Faloutsos, and Christos Faloutsos. On power-law relationships of the internet topology. *ACM SIGCOMM Computer Communication Review*, 29:251–262, 1999. DOI: 10.1145/316194.316229

[37] H. Frank and I. Frisch. Analysis and design of survivable networks. *Communication Technology, IEEE Transactions on*, 18(5):501–519, 1970. DOI: 10.1109/tcom.1970.1090419

[38] Linton C. Freeman. A set of measures of centrality based on betweenness. *Sociometry*, pages 35–41, 1977. DOI: 10.2307/3033543

[39] Linton C. Freeman. Centrality in social networks conceptual clarification. *Social Networks*, 1(3):215–239, 1978. DOI: 10.1016/0378-8733(78)90021-7

[40] Jianxi Gao, Sergey V. Buldyrev, Shlomo Havlin, and H. Eugene Stanley. Robustness of a network of networks. *Physical Review Letters*, 107(19):195701, 2011. DOI: 10.1103/physrevlett.107.195701

[41] Jianxi Gao, Sergey V. Buldyrev, H. Eugene Stanley, and Shlomo Havlin. Networks formed from interdependent networks. *Nature Physics*, 8(1):40–48, 2012. DOI: 10.1038/nphys2180

[42] Palash Goyal, Sujit Rokka Chhetri, and Arquimedes Canedo. dyngraph2vec: Capturing network dynamics using dynamic graph representation learning. *Knowledge-Based Systems*, 187:104816, 2020. DOI: 10.1016/j.knosys.2019.06.024

[43] Palash Goyal, Nitin Kamra, Xinran He, and Yan Liu. Dyngem: Deep embedding method for dynamic graphs. *ArXiv Preprint ArXiv:1805.11273*, 2018.

[44] Frank Harary and Allen Schwenk. The spectral approach to determining the number of walks in a graph. *Pacific Journal of Mathematics*, 80(2):443–449, 1979. DOI: 10.2140/pjm.1979.80.443

[45] Lenwood S. Heath and Allan A. Sioson. Multimodal networks: Structure and operations. *Computational Biology and Bioinformatics, IEEE/ACM Transactions on*, 6(2):321–332, 2009. DOI: 10.1109/tcbb.2007.70243

[46] Shlomo Hoory, Nathan Linial, and Avi Wigderson. Expander graphs and their applications. *Bulletin of the American Mathematical Society*, 43(4):439–561, 2006. DOI: 10.1090/s0273-0979-06-01126-8

[47] Yifan Hu, Yehuda Koren, and Chris Volinsky. Collaborative filtering for implicit feedback datasets. *Data Mining, ICDM. 8th IEEE International Conference on*, pages 263–272, 2008. DOI: 10.1109/icdm.2008.22

[48] Xuqing Huang, Jianxi Gao, Sergey V. Buldyrev, Shlomo Havlin, and H. Eugene Stanley. Robustness of interdependent networks under targeted attack. *Physical Review E*, 83(6):065101, 2011. DOI: 10.1103/physreve.83.065101

[49] Mark Jerrum and Alistair Sinclair. Conductance and the rapid mixing property for Markov chains: The approximation of permanent resolved. *Proc. of the 20th Annual ACM Symposium on Theory of Computing*, pages 235–244, 1988. DOI: 10.1145/62212.62234

[50] Wei Jin, Yaxing Li, Han Xu, Yiqi Wang, Shuiwang Ji, Charu Aggarwal, and Jiliang Tang. Adversarial attacks and defenses on graphs. *ACM SIGKDD Explorations Newsletter*, 22(2):19–34, 2021. DOI: 10.1145/3447556.3447566

[51] WU Jun, Mauricio Barahona, Tan Yue-Jin, and Deng Hong-Zhong. Natural connectivity of complex networks. *Chinese Physics Letters*, 27(7):078902, 2010. DOI: 10.1088/0256-307x/27/7/078902

[52] Seyed Mehran Kazemi, Rishab Goel, Kshitij Jain, Ivan Kobyzev, Akshay Sethi, Peter Forsyth, and Pascal Poupart. Representation learning for dynamic graphs: A survey. *Journal of Machine Learning Research*, 21(70):1–73, 2020.

[53] David Kempe, Jon Kleinberg, and Éva Tardos. Maximizing the spread of influence through a social network. *Proc. of the 9th ACM SIGKDD International Conference on Knowledge Discovery and Data Mining*, pages 137–146, 2003. DOI: 10.1145/956750.956769

[54] Samir Khuller, Anna Moss, and Joseph Seffi Naor. The budgeted maximum coverage problem. *Information Processing Letters*, 70(1):39–45, 1999. DOI: 10.1016/s0020-0190(99)00031-9

[55] Mikko Kivelä, Alex Arenas, Marc Barthelemy, James P. Gleeson, Yamir Moreno, and Mason A. Porter. Multilayer networks. *Journal of Complex Networks*, 2(3):203–271, 2014. DOI: 10.1093/comnet/cnu016

[56] Jon Kleinberg and Eva Tardos. *Algorithm Design*. Pearson Education India, 2006.

[57] Jon M. Kleinberg. Authoritative sources in a hyperlinked environment. *ACM-SIAM Symposium on Discrete Algorithms*, 1998. DOI: 10.1145/324133.324140

[58] Michael A. Kohanski, Daniel J. Dwyer, and James J. Collins. How antibiotics kill bacteria: From targets to networks. *Nature Reviews Microbiology*, 8(6):423, 2010. DOI: 10.1038/nrmicro2333

[59] Yehuda Koren, Robert Bell, and Chris Volinsky. Matrix factorization techniques for recommender systems. *Computer*, (8):30–37, 2009. DOI: 10.1109/mc.2009.263

[60] Istvan A. Kovacs and Albert-Laszlo Barabasi. Network science: Destruction perfected. *Nature*, 524(7563):38–39, 2015. DOI: 10.1038/524038a

[61] Renaud Lambiotte, Martin Rosvall, and Ingo Scholtes. Understanding complex systems: From networks to optimal higher-order models. *ArXiv Preprint ArXiv:1806.05977*, 2018.

[62] Long T. Le, Tina Eliassi-Rad, and Hanghang Tong. Met: A fast algorithm for minimizing propagation in large graphs with small eigen-gaps. *Proc. of the SIAM International Conference on Data Mining*, pages 694–702, 2015. DOI: 10.1137/1.9781611974010.78

[63] Daniel D. Lee and H. Sebastian Seung. Algorithms for non-negative matrix factorization. *Advances in Neural Information Processing Systems*, pages 556–562, 2001. DOI: 10.1145/1232722.1232727

[64] Jure Leskovec, Lada A. Adamic, and Bernardo A. Huberman. The dynamics of viral marketing. *ACM Transactions on the Web (TWEB)*, 1(1):5, 2007.

[65] Jure Leskovec, Jon Kleinberg, and Christos Faloutsos. Graphs over time: Densification laws, shrinking diameters and possible explanations. *Proc. of the 11th ACM SIGKDD International Conference on Knowledge Discovery in Data Mining*, pages 177–187, 2005. DOI: 10.1145/1081870.1081893

[66] Jure Leskovec, Jon Kleinberg, and Christos Faloutsos. Graph evolution: Densification and shrinking diameters. *ACM Transactions on Knowledge Discovery from Data (TKDD)*, 1(1):2, 2007. DOI: 10.1145/1217299.1217301

[67] Jure Leskovec, Andreas Krause, Carlos Guestrin, Christos Faloutsos, Jeanne VanBriesen, and Natalie Glance. Cost-effective outbreak detection in networks. *Proc. of the 13th ACM SIGKDD International Conference on Knowledge Discovery and Data Mining*, pages 420–429, 2007. DOI: 10.1145/1281192.1281239

[68] Keith Levin, Fred Roosta, Michael Mahoney, and Carey Priebe. Out-of-sample extension of graph adjacency spectral embedding. *International Conference on Machine Learning*, pages 2975–2984, PMLR, 2018.

[69] Bin Li, Qiang Yang, and Xiangyang Xue. Can movies and books collaborate? Cross-domain collaborative filtering for sparsity reduction. *IJCAI*, 9:2052–2057, 2009.

[70] Jundong Li, Harsh Dani, Xia Hu, Jiliang Tang, Yi Chang, and Huan Liu. Attributed network embedding for learning in a dynamic environment. *Proc. of the ACM on Conference on Information and Knowledge Management*, pages 387–396, 2017. DOI: 10.1145/3132847.3132919

[71] Jundong Li, Xia Hu, Liang Wu, and Huan Liu. Robust unsupervised feature selection on networked data. *Proc. of the SIAM International Conference on Data Mining*, pages 387–395, 2016. DOI: 10.1137/1.9781611974348.44

[72] Liangyue Li, Hanghang Tong, Nan Cao, Kate Ehrlich, Yu-Ru Lin, and Norbou Buchler. Replacing the irreplaceable: Fast algorithms for team member recommendation. *Proc. of the 24th International Conference on World Wide Web*, pages 636–646, ACM, 2015. DOI: 10.1145/2736277.2741132

[73] Liangyue Li, Hanghang Tong, Yanghua Xiao, and Wei Fan. Cheetah: Fast graph kernel tracking on dynamic graphs. *SDM*, SIAM, 2015. DOI: 10.1137/1.9781611974010.32

[74] Rong-Hua Li and Jeffrey Xu Yu. Triangle minimization in large networks. *Knowledge and Information Systems*, 45(3):617–643, 2015. DOI: 10.1007/s10115-014-0800-9

[75] Yanen Li, Jia Hu, ChengXiang Zhai, and Ye Chen. Improving one-class collaborative filtering by incorporating rich user information. *Proc. of the 19th ACM International Conference on Information and Knowledge Management*, pages 959–968, 2010. DOI: 10.1145/1871437.1871559

[76] Chuan-bi Lin. Projected gradient methods for nonnegative matrix factorization. *Neural Computation*, 19(10):2756–2779, 2007. DOI: 10.1162/neco.2007.19.10.2756

[77] Jialu Liu, Chi Wang, Jing Gao, Quanquan Gu, Charu C. Aggarwal, Lance M. Kaplan, and Jiawei Han. Gin: A clustering model for capturing dual heterogeneity in networked data. *SDM*, pages 388–396, SIAM, 2015. DOI: 10.1137/1.9781611974010.44

[78] Zhongqi Lu, Weike Pan, Evan Wei Xiang, Qiang Yang, Lili Zhao, and ErHeng Zhong. Selective transfer learning for cross domain recommendation. *SDM*, pages 641–649, SIAM, 2013. DOI: 10.1137/1.9781611972832.71

[79] Hao Ma, Dengyong Zhou, Chao Liu, Michael R. Lyu, and Irwin King. Recommender systems with social regularization. *Proc. of the 4th ACM International Conference on Web Search and Data Mining*, pages 287–296, 2011. DOI: 10.1145/1935826.1935877

[80] Sedigheh Mahdavi, Shima Khoshraftar, and Aijun An. dynnode2vec: Scalable dynamic network embedding. *IEEE International Conference on Big Data (Big Data)*, pages 3762–3765, 2018. DOI: 10.1109/bigdata.2018.8621910

[81] Franco Manessi, Alessandro Rozza, and Mario Manzo. Dynamic graph convolutional networks. *Pattern Recognition*, 97:107000, 2020. DOI: 10.1016/j.patcog.2019.107000

[82] Julian Mcauley and Jure Leskovec. Discovering social circles in ego networks *ACM Transactions on Knowledge Discovery from Data (TKDD)*, 8(1):4, 2014. DOI 10.1145/2556612

[83] Ron Milo, Shai Shen-Orr, Shalev Itzkovitz, Nadav Kashtan, Dmitri Chklovskii, and Uri Alon. Network motifs: Simple building blocks of complex networks. *Science* 298(5594):824–827, 2002. DOI: 10.1126/science.298.5594.824

[84] Sudatta Mohanty and Alexey Pozdnukhov. Graph CNN+ LSTM framework for dynamic macroscopic traffic congestion prediction. *International Workshop on Mining and Learning with Graphs*, 2018.

[85] James Moody and Douglas R. White. Social cohesion and embeddedness: A hierarchic conception of social groups. *American Sociological Review*, pages 1–25, 2003.

[86] Flaviano Morone and Hernán A. Makse. Influence maximization in complex networks through optimal percolation. *Nature*, 524(7563):65, 2015. DOI: 10.1038/nature1460

[87] Apurva Narayan and Peter H. O'N Roe. Learning graph dynamics using deep neural networks. *IFAC-PapersOnLine*, 51(2):433–438, 2018. DOI: 10.1016/j.ifacol.2018.03.074

[88] George L. Nemhauser, Laurence A. Wolsey, and Marshall L. Fisher. An analysis of approximations for maximizing submodular set functions—I. *Mathematical Programming*, 14(1):265–294, 1978. DOI: 10.1007/bf01588971

[89] Mark E. J. Newman. A measure of betweenness centrality based on random walks. *Social Networks*, 27(1):39–54, 2005. DOI: 10.1016/j.socnet.2004.11.009

[90] Mark E. J. Newman. The mathematics of networks. *The New Palgrave Encyclopedia of Economics*, 2(2008):1–12, 2008.

[91] Dung T. Nguyen, Yilin Shen, and My T. Thai. Detecting critical nodes in interdependent power networks for vulnerability assessment. *IEEE Transactions on Smart Grid*, 4(1):151–159, 2013. DOI: 10.1109/tsg.2012.2229398

[92] Jingchao Ni, Hanghang Tong, Wei Fan, and Xiang Zhang. Inside the atoms: Ranking on a network of networks. *Proc. of the 20th ACM SIGKDD International Conference on Knowledge Discovery and Data Mining*, pages 1356–1365, 2014. DOI: 10.1145/2623330.2623643

[93] Huazhong Ning, Wei Xu, Yun Chi, Yihong Gong, and Thomas S. Huang. Incremental spectral clustering by efficiently updating the eigen-system. *Pattern Recognition*, 43(1):113–127, 2010. DOI: 10.1016/j.patcog.2009.06.001

[94] Lawrence Page, Sergey Brin, Rajeev Motwani, and Terry Winograd. The PageRank citation ranking: Bringing order to the Web. *Technical Report*, Stanford Digital Library Technologies Project, 1998. Paper SIDL-WP-1999-0120 (version of 11/11/1999).

[95] Rong Pan, Yunhong Zhou, Bin Cao, Nathan N. Liu, Rajan Lukose, Martin Scholz, and Qiang Yang. One-class collaborative filtering. *Data Mining, ICDM. 8th IEEE International Conference on*, pages 502–511, 2008. DOI: 10.1109/icdm.2008.16

[96] Roni Parshani, Sergey V. Buldyrev, and Shlomo Havlin. Interdependent networks: Reducing the coupling strength leads to a change from a first to second order percolation transition. *Physical Review Letters*, 105(4):048701, 2010. DOI: 10.1103/physrevlett.105.048701

[97] B. Aditya Prakash, Deepayan Chakrabarti, Nicholas C. Valler, Michalis Faloutsos, and Christos Faloutsos. Threshold conditions for arbitrary cascade models on arbitrary networks. *Knowledge and Information Systems*, 33(3):549–575, 2012. DOI: 10.1007/s10115-012-0520-y

[98] B. Aditya Prakash, Ashwin Sridharan, Mukund Seshadri, Sridhar Machiraju, and Christos Faloutsos. Eigenspokes: Surprising patterns and scalable community chipping in large graphs. *Advances in Knowledge Discovery and Data Mining, 14th Pacific-Asia Conference, PAKDD, Proceedings. Part II*, pages 435–448, Hyderabad, India, June 21–24, 2010. DOI: 10.1007/978-3-642-13672-6_42

[99] Mahmudur Rahman and Mohammad Al Hasan. Link prediction in dynamic networks using graphlet. *Joint European Conference on Machine Learning and Knowledge Discovery in Databases*, pages 394–409, Springer, 2016. DOI: 10.1007/978-3-319-46128-1_25

[100] Sabry Razick, George Magklaras, and Ian M. Donaldson. irefindex: A consolidated protein interaction database with provenance. *BMC Bioinformatics*, 9(1):1, 2008. DOI: 10.1186/1471-2105-9-405

[101] Steven M. Rinaldi, James P. Peerenboom, and Terrence K. Kelly. Identifying, understanding, and analyzing critical infrastructure interdependencies. *Control Systems, IEEE*, 21(6):11–25, 2001. DOI: 10.1109/37.969131

[102] Vittorio Rosato, L. Issacharoff, F. Tiriticco, Sandro Meloni, S. Porcellinis, and Roberto Setola. Modelling interdependent infrastructures using interacting dynamical models. *International Journal of Critical Infrastructures*, 4(1–2):63–79, 2008. DOI: 10.1504/ijcis.2008.016092

[103] Martin Rosvall, Alcides V. Esquivel, Andrea Lancichinetti, Jevin D. West, and Renaud Lambiotte. Memory in network flows and its effects on spreading dynamics and community detection. *Nature Communications*, 5(1):1–13, 2014. DOI: 10.1038/ncomms5630

[104] Mandana Saebi, Jian Xu, Lance M. Kaplan, Bruno Ribeiro, and Nitesh V. Chawla. Efficient modeling of higher-order dependencies in networks: From algorithm to application for anomaly detection. *EPJ Data Science*, 9(1):15, 2020. DOI: 10.1140/epjds/s13688-020-00233-y

[105] Hooman Peiro Sajjad, Andrew Docherty, and Yuriy Tyshetskiy. Efficient representation learning using random walks for dynamic graphs. *ArXiv Preprint ArXiv:1901.01346*, 2019.

[106] Rubén J. Sánchez-García, Emanuele Cozzo, and Yamir Moreno. Dimensionality reduction and spectral properties of multilayer networks. *Physical Review E*, 89(5):052815, 2014. DOI: 10.1103/physreve.89.052815

[107] Ingo Scholtes. When is a network a network? Multi-order graphical model selection in pathways and temporal networks. *Proc. of the 23rd ACM SIGKDD International Conference on Knowledge Discovery and Data Mining*, pages 1037–1046, 2017. DOI: 10.1145/3097983.3098145

[108] Arunabha Sen, Anisha Mazumder, Joydeep Banerjee, Arun Das, and Randy Compton. Multi-layered network using a new model of interdependency. *ArXiv Preprint ArXiv:1401.1783*, 2014.

[109] Youngjoo Seo, Michaël Defferrard, Pierre Vandergheynst, and Xavier Bresson. Structured sequence modeling with graph convolutional recurrent networks. *International Conference on Neural Information Processing*, pages 362–373, Springer, 2018. DOI: 10.1007/978-3-030-04167-0_33

[110] Jia Shao, Sergey V. Buldyrev, Shlomo Havlin, and H. Eugene Stanley. Cascade of failures in coupled network systems with multiple support-dependence relations. *Physical Review E*, 83(3):036116, 2011. DOI: 10.1103/physreve.83.036116

[111] Yue Shi, Alexandros Karatzoglou, Linas Baltrunas, Martha Larson, Nuria Oliver, and Alan Hanjalic. CLiMF: Learning to maximize reciprocal rank with collaborative less-is-more filtering. *Proc. of the 6th ACM Conference on Recommender Systems*, pages 139–146, 2012. DOI: 10.1145/2365952.2365981

[112] Ajit P. Singh and Geoffrey J. Gordon. Relational learning via collective matrix factorization. *Proc. of the 14th ACM SIGKDD International Conference on Knowledge Discovery and Data Mining*, pages 650–658, 2008. DOI: 10.1145/1401890.1401969

[113] Michael Sipser. *Introduction to the Theory of Computation*. PWS Publishing Company, 1997. DOI: 10.1145/230514.571645

[114] G. W. Stewart and Ji-Guang Sun. *Matrix Perturbation Theory*. Academic Press, 1990.

[115] Lichao Sun, Yingtong Dou, Carl Yang, Ji Wang, Philip S. Yu, Lifang He, and Bo Li. Adversarial attack and defense on graph data: A survey. *ArXiv Preprint ArXiv:1812.10528*, 2018.

[116] Yiwei Sun, Suhang Wang, Xianfeng Tang, Tsung-Yu Hsieh, and Vasant Honavar. Node injection attacks on graphs via reinforcement learning. *ArXiv Preprint ArXiv:1909.06543*, 2019.

[117] Jie Tang, Jing Zhang, Limin Yao, Juanzi Li, Li Zhang, and Zhong Su. ArnetMiner: Extraction and mining of academic social networks. *Proc. of the 14th ACM SIGKDD International Conference on Knowledge Discovery and Data Mining*, pages 990–998, 2008. DOI: 10.1145/1401890.1402008

[118] Jiliang Tang, Huiji Gao, and Huan Liu. mTrust: Discerning multi-faceted trust in a connected world. *Proc. of the 5th ACM International Conference on Web Search and Data Mining*, pages 93–102, 2012. DOI: 10.1145/2124295.2124309

148 BIBLIOGRAPHY

[119] Jiliang Tang, Huiji Gao, Huan Liu, and Atish Das Sarma. eTrust: Understanding trust evolution in an online world. *Proc. of the 18th ACM SIGKDD International Conference on Knowledge Discovery and Data Mining*, pages 253–261, 2012. DOI: 10.1145/2339530.2339574

[120] Lei Tang and Huan Liu. Relational learning via latent social dimensions. *Proc. of the 15th ACM SIGKDD International Conference on Knowledge Discovery and Data Mining*, pages 817–826, 2009. DOI: 10.1145/1557019.1557109

[121] Robert Tarjan. Depth-first search and linear graph algorithms. *SIAM Journal on Computing*, 1(2):146–160, 1972. DOI: 10.1137/0201010

[122] Hanghang Tong, Spiros Papadimitriou, Philip S. Yu, and Christos Faloutsos. Fast monitoring proximity and centrality on time-evolving bipartite graphs. *Statistical Analysis and Data Mining*, 1(3):142–156, 2008. DOI: 10.1002/sam.10014

[123] Hanghang Tong, B. Aditya Prakash, Tina Eliassi-Rad, Michalis Faloutsos, and Christos Faloutsos. Gelling, and melting, large graphs by edge manipulation. *Proc. of the 21st ACM International Conference on Information and Knowledge Management*, pages 245–254, 2012. DOI: 10.1145/2396761.2396795

[124] Hanghang Tong, B. Aditya Prakash, Charalampos Tsourakakis, Tina Eliassi-Rad, Christos Faloutsos, and Duen Horng Chau. On the vulnerability of large graphs. *Data Mining (ICDM), IEEE 10th International Conference on*, pages 1091–1096, 2010. DOI: 10.1109/icdm.2010.54

[125] Charalampos E. Tsourakakis. Fast counting of triangles in large real networks without counting: Algorithms and laws. *Data Mining, ICDM. 8th IEEE International Conference on*, pages 608–617, 2008. DOI: 10.1109/icdm.2008.72

[126] Marc A. Van Driel, Jorn Bruggeman, Gert Vriend, Han G. Brunner, and Jack A. M Leunissen. A text-mining analysis of the human phenome. *European Journal of Human Genetics*, 14(5):535–542, 2006. DOI: 10.1038/sj.ejhg.5201585

[127] Alessandro Vespignani. Complex networks: The fragility of interdependency. *Nature* 464(7291):984–985, 2010. DOI: 10.1038/464984a

[128] Fei Wang, Hanghang Tong, and Ching-Yung Lin. Towards evolutionary nonnegative matrix factorization. *25th AAAI Conference on Artificial Intelligence*, 2011.

[129] Jihong Wang, Minnan Luo, Fnu Suya, Jundong Li, Zijiang Yang, and Qinghua Zhen Scalable attack on graph data by injecting vicious nodes. *Data Mining and Knowledge Discovery*, 34(5):1363–1389, 2020. DOI: 10.1007/s10618-020-00696-7

[130] Tong Wang, Xing-Sheng He, Ming-Yang Zhou, and Zhong-Qian Fu. Link prediction in evolving networks based on popularity of nodes. *Scientific Reports*, 7(1):1–10, 2017. DOI: 10.1038/s41598-017-07315-4

[131] Yang Wang, Deepayan Chakrabarti, Chenxi Wang, and Christos Faloutsos. Epidemic spreading in real networks: An eigenvalue viewpoint. *Reliable Distributed Systems. Proceedings. 22nd International Symposium on*, pages 25–34, IEEE, 2003. DOI: 10.1109/reldis.2003.1238052

[132] Stanley Wasserman. *Social Network Analysis: Methods and Applications*, vol. 8, Cambridge University Press, 1994. DOI: 10.1017/cbo9780511815478

[133] Duncan J. Watts and Steven H. Strogatz. Collective dynamics of small-world networks. *Nature*, 393(6684):440–442, 1998. DOI: 10.1038/30918

[134] V. Vassilevska Williams. Multiplying matrices faster than Coppersmith–Winograd, *Proc. of the 44th Annual ACM Symposium on Theory of Computing*, pages 887–898, 2012.

[135] Jun Wu, Barahona Mauricio, Yue-Jin Tan, and Hong-Zhong Deng. Natural connectivity of complex networks. *Chinese Physics Letters*, 27(7):78902, 2010. DOI: 10.1088/0256-307x/27/7/078902

[136] Chang Xu, Dacheng Tao, and Chao Xu. A survey on multi-view learning. *ArXiv Preprint ArXiv:1304.5634*, 2013.

[137] Jian Xu, Thanuka L. Wickramarathne, and Nitesh V. Chawla. Representing higher-order dependencies in networks. *Science Advances*, 2(5):e1600028, 2016. DOI: 10.1126/sci-adv.1600028

[138] Deqing Yang, Jingrui He, Huazheng Qin, Yanghua Xiao, and Wei Wang. A graph-based recommendation across heterogeneous domains. *Proc. of the 24rd ACM International Conference on Conference on Information and Knowledge Management*, pages 463–472, 2015. DOI: 10.1145/2806416.2806523

[139] Yuan Yao, Hanghang Tong, Guo Yan, Feng Xu, Xiang Zhang, Boleslaw K. Szymanski, and Jian Lu. Dual-regularized one-class collaborative filtering. *Proc. of the 23rd ACM International Conference on Conference on Information and Knowledge Management*, pages 759–768, 2014. DOI: 10.1145/2661829.2662042

[140] Hao Yin, Austin R. Benson, and Jure Leskovec. The local closure coefficient: A new perspective on network clustering. *Networks*, 26(41):44, 2019. DOI: 10.1145/3289600.3290991

[141] Hao Yin, Austin R. Benson, Jure Leskovec, and David F. Gleich. Local higher-order graph clustering. *Proc. of the 23rd ACM SIGKDD International Conference on Knowledge Discovery and Data Mining*, pages 555–564, 2017. DOI: 10.1145/3097983.3098069

[142] Wenchao Yu, Wei Cheng, Charu C. Aggarwal, Haifeng Chen, and Wei Wang. Link prediction with spatial and temporal consistency in dynamic networks. *IJCAI*, pages 3343–3349, 2017. DOI: 10.24963/ijcai.2017/467

[143] Ziwei Zhang, Peng Cui, Jian Pei, Xiao Wang, and Wenwu Zhu. Timers: Error-bounded SVD restart on dynamic networks. *Proc. of the AAAI Conference on Artificial Intelligence*, vol. 32, 2018.

[144] Dawei Zhou, Jingrui He, K. Seluk Candan, and Hasan Davulcu. MUVIR: Multi-view rare category detection. *Proc. of the 24th International Joint Conference on Artificial Intelligence, IJCAI*, pages 4098–4104, Buenos Aires, Argentina, July 25–31, 2015.

[145] Daniel Zügner, Amir Akbarnejad, and Stephan Günnemann. Adversarial attacks on neural networks for graph data. *Proc. of the 24th ACM SIGKDD International Conference on Knowledge Discovery and Data Mining*, pages 2847–2856, 2018. DOI: 10.1145/3219819.3220078

Authors' Biographies

CHEN CHEN

Chen Chen is currently a Research Assistant Professor at the University of Virginia. Before joining the University of Virginia, she was a software engineer at Google working on personalized recommendations for Google Assistant. Chen received her Ph.D. from Arizona State University. Her research has focused on the connectivity of complex networks, which has been applied to address pressing challenges in various high-impact domains, including social media, bioinformatics, recommendation, and critical infrastructure systems. Her research has appeared in top-tier conferences (including KDD, ICDM, SDM, WSDM, and DASFAA), and prestigious journals (including IEEE TKDE, ACM TKDD, and SIAM SAM). Chen has received several awards, including Bests of SDM'15, Bests of KDD'16, Rising Star in EECS'19, and Outstanding Reviewer of WSDM'21.

HANGHANG TONG

Hanghang Tong is currently an associate professor at the Department of Computer Science at University of Illinois at Urbana-Champaign. Before that, he was an associate professor at the School of Computing, Informatics, and Decision Systems Engineering (CIDSE), Arizona State University. He received his M.Sc. and Ph.D. from Carnegie Mellon University in 2008 and 2009, respectively, both in Machine Learning. His research interest is in large-scale data mining for graphs and multimedia. He has received several awards, including SDM/IBM Early Career Data Mining Research award (2018), NSF CAREER award (2017), ICDM 10-Year Highest Impact Paper award (2015), four best paper awards (TUP'14, CIKM'12, SDM'08, ICDM'06), seven "bests of conference," one best demo, honorable mention (SIGMOD'17), and one best demo candidate, second place (CIKM'17). He has published over 100 refereed articles. He is the Editor-in-Chief of *SIGKDD Explorations* (ACM), an action editor of *Data Mining and Knowledge Discovery* (Springer), and an associate editor of *Knowledge and Information Systems* (Springer) and *Neurocomputing Journal* (Elsevier). He has served as a program committee member in multiple data mining, database, and artificial intelligence venues (e.g., SIGKDD, SIGMOD, AAAI, WWW, CIKM, etc.).

Printed in the United States
by Baker & Taylor Publisher Services